PLANT ECOLOGY - TRADITIONAL APPROACHES TO RECENT TRENDS

Edited by **Zubaida Yousaf**

Plant Ecology - Traditional Approaches to Recent Trends
http://dx.doi.org/10.5772/65821
Edited by Zubaida Yousaf

Contributors

Toshihiro Yoshihara, Allan Sebata, Germani Concenco, Patricia Menegaz De Farias, André Andres, Fábio Schreiber, Michely Tomazi, Leandro Galon, Néstor Fabio Alzate-Quintero, Ivana Moisinho, Mariane Camponogara Coradini, William Ceolin, Özlem Aksoy, Renato De Mello Prado, Gilmara Pereira Da Silva, Chieh-Chen Huang, Ying-Ning Ho, Dony Chacko Mathew, Jimin Cheng, Chengcheng Gang, Liang Guo, Wei Li, Jingwei Jin, Jishuai Su, Lin Wei, Eusebio Cano Carmona, Cano Ortiz Ana, Musarella Carmelo María, Zubaida Yousaf

Published by InTech
Janeza Trdine 9, 51000 Rijeka, Croatia

Publishing Process Manager Martina Usljebrka
Technical Editor SPi Global
Cover InTech Design team

First published August, 2017
Printed in Croatia
Legal deposit, Croatia: National and University Library in Zagreb

Additional hard copies can be obtained from orders@intechopen.com

Plant Ecology - Traditional Approaches to Recent Trends, Edited by Zubaida Yousaf
p. cm.
Print ISBN 978-953-51-3339-1
Online ISBN 978-953-51-3340-7

For EU product safety concerns:
IN TECH d.o.o., Prolaz Marije Krucifikse Kozulić 3, 51000 Rijeka, Croatia,
info@intechopen.com or visit our website at intechopen.com.

Meet the editor

Dr. Zubaida Yousaf is working as an associate professor in the Department of Botany, Lahore College for Women University, Lahore. She joined this institute in 2009 as an assistant professor. She did her postdoc from South China Botanical Garden, Guangzhou, China, in 2011, funded by TWAS-CAS. She has authored 45 research articles and 4 books and contributed 6 chapters in international editors' book. She has supervised more than 40 MS thesis, and currently 5 PhD students are working under her supervision.

Contents

Preface IX

Section 1 Plant Ecology 1

Chapter 1 **Introductory Chapter: Plant Ecology 3**
Zubaida Yousaf and Habiba Ramazan

Section 2 Phylogenetics Ecology 7

Chapter 2 **Detection of Environmental Mutagens Through Plant Bioassays 9**
Özlem Aksoy

Chapter 3 **Ecology of Woody Plants in African Savanna Ecosystems 25**
Allan Sebata

Chapter 4 **Modification in Grassland Ecology under the Influence of Changing Climatic and Land Use Conditions 43**
Jimin Cheng, Chengcheng Gang, Liang Guo, Wei Li, Jingwei Jin, Jishuai Su and Lin Wei

Chapter 5 **Leaf Ecology and Radiocesium Contamination in Trees/Forests 65**
Toshihiro Yoshihara

Section 3 Functional Ecology 91

Chapter 6 **Plant-Microbe Ecology: Interactions of Plants and Symbiotic Microbial Communities 93**
Ying-Ning Ho, Dony Chacko Mathew and Chieh-Chen Huang

Chapter 7 **Phytosociological Surveys in Weed Science: Old Concept, New Approach 121**
Germani Concenço, Patrícia Menegaz de Farias, Néstor Fabio Alzate Quintero, Fábio Schreiber, Leandro Galon, Michely Tomazi, Ivana Santos Moisinho, Mariane Camponogara Coradini, William Christofari Ceolin and André Andres

Chapter 8 **Ecological Response to Global Change: Changes in C:N:P Stoichiometry in Environmental Adaptations of Plants 147**
Renato de Mello Prado and Gilmara Pereira da Silva

Chapter 9 **Biogeographical Areas of Hispaniola (Dominican Republic, Republic of Haiti) 165**
Ana Cano Ortiz, Carmelo María Musarella and Eusebio Cano Carmona

Preface

A galactic literature could be found on the captivating topic plant ecology, but the apprentices on this wide topic could never be conclusive. This book is aimed to cover the phylogenetic and functional ecology with special reference to ecological shifts. I hope this book may benefit the students, fellow professors, and resource managers studying plant sciences. Since the topics stated in this book are not new but the issues and technologies mentioned were new to me, I expect that they will be new and equally advanced for the readers too.

While editing this book, I kept few key assumptions relevant to the experience of my audience and available information about life sciences. Hence, I deliberately wrote for the audiences who are already well familiarized with plants and ecology. The key element included into the text of this book covers:

[1] The categorization of plant populations and communities within the ecosystem
[2] The relationship of plants with their micro- and macrohabitat
[3] The relationship and competition of plants within the same group and community
[4] The positioning of plants and plant communities after the alterations in biosphere
[5] General models and their applications for the theoretical and applied conceptualization

I encourage the readers to get out into the field to identify plants and to dig out the anthropogenic and social activities effecting plants to come along with the development of plant ecology; to rise and serve the topic of the enormous number of plants facing extinction; and to relish themselves and make some effort to contribute something to the world.

I want to thank InTech Publishers for providing me such ravishing platform to work with the young and emerging scientists. It has been a great experience working with them and interacting with international scientists. Also, I would also want to thank the Lahore College for Women University for cherishing and encouraging its employers and students to cater their contributions in the field of life sciences.

Dr. Zubaida Yousaf
Lahore College for Women University,
Lahore, Pakistan

Section 1

Plant Ecology

Chapter 1

Introductory Chapter: Plant Ecology

Zubaida Yousaf and Habiba Ramazan

Additional information is available at the end of the chapter

http://dx.doi.org/10.5772/intechopen.69926

Plants are one of the densest ecotypes with vicariate diversity in their life history depending on their mores. The ecology of land entirely depends upon plants as they are practical pioneers of the planet earth. Initiating from microorganisms to the macroorganism, they meet their living requisites expending plants. Admitting this, different vegetation types, such as grassland and timberland, are major biomes which help to flux the environmental shifts under global impact. The day-by-day advances in science and technology and ever-emerging needs of living organisms are some anthropogenic activities which had entirely changed the scenario of green systems of the world. Such drastic changes in plant ecology are harmful for all the beings and other respective factors. These substantial amendments occurred due to the earthquakes, industrial effluents, forest fires, carbon ignition, destruction of vegetative and agronomic landforms, etc.

1. Traditional ecology

Broadly stating the relationship of plants with the other living organisms and environmental factors is called plant ecology [1]. But elaborating the term, plant ecology is the study of the relationship of plants with the biotic (living organisms such as animals and other plants, bacteria, and fungi) and abiotic factors such as moisture, temperature, sunlight, soil (nutrients and salinity), and water (rainfall and water table) surrounding them. By the passage of time, the addressed issues regarding biosphere came into consideration. Though from the time of Alexander Von Humboldt (father of ecology) the known species of plants were about 20,000, now the number increased up to 40,0000, but the changes in elements of biosphere are increasing the issues such as loss of habitat, plant, animal, microflora, mutation, pollution, and soil sickness [2–4]. Due to these issues, the most affected living organisms are plants which urged the scientists to investigate the root causes of such drastic changes and commotion to the plant ecology. According to the climate, human and animal interaction, flora, and fauna, the planet earth is categorized into biomes. There are about six major biomes with cutting clarity of subcategories. The largest biome is boreal/coniferous forest; however, the second largest biome is grasslands that are ubiquitous as compared to other biomes; tropical rainforest covers only 6% of the world, but they have the richest biodiversity; however, the hottest biome is desert with the minimal biodiversity; in contrast the coldest biome is alpine forest merely with considerable biodiversity [5]. Specifically, the plant populations have dominantly occupied this globe; according to an estimation, 99.9% area of planet earth is covered with flora [6]. About 350,000 species of plants excluding ferns, bryophytes, and algae are known and documented yet. Among them approximately 20% are under the risk of endangerment [7]. The risk of endangerment or extinction due to natural and unnatural disasters has disturbed the whole food chain and web and is continually pushing toward the worst conditions [8].

2. Phylogenetic ecology

Whenever ecological drift and loss in biodiversity of living organisms are discussed, it is generally apprehended that plants are vanishing due to overgrazing and animals are dying due to the inaccessibility of plants. But this whole globe is alive and functional on a single principle named balanced metabolic dynamics ratio between autotrophs (plants, producers) and heterotrophs (animals and microbes, consumers and decomposer) [9]. Basically, this trophic dynamics between producer and consumer within the biosphere is regulated by the transfer of energy from one part of the ecosystem to the other and even within the same ecosystem also known as energy flow in ecosystem. Except solar radiation (external source of energy), all the other energy systems are recycled and balance the dynamics of trophic level followed by complex metabolic mechanisms within the biosphere [10].

Drafting the origin of plant, their functional types and phylogenetic/evolutionary patterns are the most needed steps to timely track and record the drifts and risks to the ecosystem and biosphere [11–14]. As the current dynamics, composition and distribution of plants are altered thence, elaborating and redefining the relationship of plants with the factors encompassing them had led the flora and environment on the verge of endangerment are also expatiated, and many successful solutions to indemnify these issues are contributed by the scientists. Sustaining to this several concepts such as phylogeny, phenology, phytosociology, physiology, and anatomy of plants were used for modeling and surveying [15].

Terrestrial vegetation plays a phenomenal role in management of landscape and hydrological regime. Also the climatic change can be ameliorated by them as they could better regulate biogeological water cycle and sequestrate carbon cycle [16]. The provision of protection against water resources by surface runoff leading toward flood attenuation, aquifer recharging, sea water leveling, water table leveling and fresh water management. Increase in temperature, variation in precipitation, and extreme events have potentially manifested the natural conversational and agricultural management regimes including an indirect risk that was constrained for social and human livelihoods [16, 17]. Although most of the commotion was inflicted due to water regime mismanagement, actually the fraction has been completed by burning practice, by grazing, or by harvesting hay/fodder (directly and indirectly, respectively). Such climatic changes and management conflict of water and vegetation regimes grounded the grasslands toward threats which were also highlighted by Intergovernmental Panel on Climate Change (IPCC) 2014 in their report [18]. Hence, the study of plant ecology is a fountainhead step toward the investigation of cause and solution of biological metabolism and their functioning in the biosphere.

Author details

Zubaida Yousaf* and Habiba Ramazan

*Address all correspondence to: z4zubaida@yahoo.com

Lahore College for Women University Lahore, Lahore, Pakistan

References

[1] Noble JC, Bell AD, Harper JL. The population biology of plants with clonal growth: I. The morphology and structural demography of *Carex arenaria*. The Journal of Ecology. 1979;**67**(3):983-1008

[2] McIntosh RP. The Background of Ecology: Concept and Theory. Cambridge University Press, UK; 1986

[3] Harper JL. Plant demography and ecological theory. Oikos 1980;**35**:244-253

[4] Bramwell A. Ecology in the twentieth century: A history. Bulletin of Science, Technology and Society, 1990;**10**(2):118-119

[5] Willis KJ. Evolution and function of earth's biomes: Temperate forests. In: Earth System: History and Natural Variability, UK. Vol. **III**. 2009. p. 164

[6] Rafferty JP, editor. Biomes and Ecosystems. Britannica Educational Publishing, USA; 2010

[7] Keddy P. Plants and Vegetation: Origins, Processes, Consequences. Cambridge University Press, UK; 2007.

[8] Melillo JM, McGuire AD, Kicklighter DW, Moore B, Vorosmarty CJ, Schloss AL. Global climate change and terrestrial net primary production. Nature. 1993;**363**(6426):234-240

[9] Brown JH, Gillooly JF, Allen AP, Van Savage M, West GB. Toward a metabolic theory of ecology. Ecology. 2004;**85**(7):1771-1789

[10] Lindeman RL. The trophic-dynamic aspect of ecology. Ecology 1942;**23**(4):399-417

[11] Nilan RA. Potential of plant genetic systems for monitoring and screening mutagens. Environmental Health Perspectives 1978;**27**:181

[12] Songstad DD, Petolino JF, Voytas DF & Reichert NA (2017). Genome Editing of Plants. Critical Reviews in Plant Sciences, 1-23. DOI:10.1080/07352689.2017.1281663 http://dx.doi.org/10.1080/07352689.2017.1281663

[13] Crutsinger GM,. Collins MD, Fordyce JA, Gompert Z, Nice CC, Sanders NJ. Plant genotypic diversity predicts community structure and governs an ecosystem process. Science 2006;**313**(5789):966-968

[14] Diaz S, Cabido M. Vive la difference: Plant functional diversity matters to ecosystem processes. Trends in Ecology & Evolution 2001;**16**(11):646-655

[15] Bellard C, Bertelsmeier C, Leadley P, Thuiller W, Courchamp F. Impacts of climate change on the future of biodiversity. Ecology Letters 2012;**15**(4):365-377

[16] Joyce CB, Simpson M, Casanova M. Future wet grasslands: Ecological implications of climate change. Ecosystem Health and Sustainability. **2**(9):e01240. 10.1002/ehs2.1240

[17] Parmesan, C. Ecological and evolutionary responses to recent climate change. Annual Review of Ecology, Evolution, and Systematics. 2006;**37**:637-669

[18] Parmesan C, Yohe G. A globally coherent fingerprint of climate change impacts across natural systems. Nature. 2003;**421**(6918):37-42

Section 2

Phylogenetics Ecology

Chapter 2

Detection of Environmental Mutagens Through Plant Bioassays

Özlem Aksoy

Additional information is available at the end of the chapter

http://dx.doi.org/10.5772/intechopen.69274

Abstract

Plants are present in almost all areas of the world and can accumulate many chemical compounds present in the soil, water, and atmosphere. As these chemicals which are potentially mutagenic or carcinogenic are absorbed by the plants sharing the same environment with us, bioassays on plants can be used to detect the presence of environmental hazards. Another reason for selecting plants for assessing adverse effects of these chemicals is the ease of experimentation with plants. Evaluating the effect of a substance on basic plant characteristics such as growth, survival, or reproduction is straightforward and repeatable. Thus, various plant species are commonly utilized as indicators of adverse environmental conditions. This chapter covers the detection of environmental mutagens through plant bioassays, considering the increasing importance of biomonitoring using plants for assessing the mutagenicity of relevant chemicals and industrial waste. From this point of view, a detailed literature search was made on the subject. The genotoxic, cytotoxic, and molecular studies have been investigated and the most useful and important parts and key points of these methods were summarized. This review would be useful for scientists who are planning to conduct research on plant bioassays with different types of methods and chemicals.

Keywords: ecotoxicology, cytotoxicity, genotoxicity, phytotoxicity, mutagens

1. Introduction

Plants are the essential elements of agriculture and forestry and maintain the healthy environment for the rest of the species by producing oxygen and organic carbon compounds. Higher plants are preeminent indicators of genotoxic effects caused by chemical substances existing in the environment and therefore be utilized for detecting environmental mutagens [1]. They are exposed to many stress factors including chemical compounds and radiation affecting their seed germination, seedling growth, and floral and fruit development. These stress factors can adversely affect the quality and quantity of the product with leading to morphological, anatomical, physiological, biochemical, and molecular damage to plants [2]. There are different kinds of methods for examining phytotoxicity and genotoxicity because usually there is no standard national procedure. Therefore, the parameters of these methods vary depending on the test substances, the test plants, or the individual procedures. Because of its simplicity, low cost, and relatively high sensitivity, application of plant bioassays is usually favored over other available systems in discovering adverse effects caused by chemical substances, or pollution, existing in the environment [3]. Despite these benefits described above, there are also some limitations in using plant bioassays, such as the longer life span of plants than *Escherichia coli* T. Escherich, *Salmonella typhimurium* Lignieres, *Saccharomyces cerevisiae* Meyen ex E.C. Hansen, or *Drosophila melanogaster* Meigen; likewise, there are differences between the biochemistry of plants and mammals. Nevertheless, positive correlation results have been observed between plant and mammalian systems in many reports, supporting the preference of plant bioassays in these studies [4]. Hence, plant bioassays are commonly used for screening and monitoring environmental chemicals with mutagenic and carcinogenic potential [5, 6]. The International Program on Chemical Safety (IPCS) makes and supports research programs all around the world and develops methodologies for chemical exposure [4, 7]. Many laboratories from diverse regions of the world have been sponsored by IPCS and participated in evaluating the utility of several plant bioassays for detecting the mutagenicity of environmental chemicals [8]. By means of these studies, many methods were developed to assess toxicity in plants. Some of the recent studies with plant bioassays can be seen in **Table 1**.

Plant bioassays are usually based on the detection of chromosomal abnormalities in mitosis, sister chromatid exchanges (SCEs), and, recently, on the DNA damage analysis. Point mutations

Plant species	Test substance	Method	Reference
Vicia faba L.	Wastewater	Micronucleus method	Liu et al. [9]
Tradescantia pallida (Rose) D.R.Hunt var. *purpurea*	Pesticide	Micronucleus and stamen hair bioassays	Fadic et al. [10]
Triticum aestivum L.	Aniline	Micronucleus, mitotic index, and chromosomal aberration	Tao et al. [11]
Vicia faba L.	Insecticide	Sister chromatid exchange	Quintana et al. [12]
Oryza sativa L. var *nipponbare*	Mercury	Real-time PCR FISH	Zhen et al. [13]
Capsicum baccatum L. var. *pendulum*	Ionizing radiation	TUNEL test	Scaldaferro et al. [14]
Epipremnum aureum (Linden & André) G.S.Bunting	Volatile organic compounds	Comet assay	Naroi-et et al. [15]
Acalypha indica L.	Lead stress	RAPD-PCR	Venkatachalam et al. [16]

Table 1. Some of the recent studies with plant bioassays.

such as chlorophyll mutations in leaves, waxy mutations, or embryo mutations of *Arabidopsis* are the other detection methods [17]. Seed germination, root elongation, EC50 (the concentration that lowers %50 of the root length) determination, mitotic index, chromosomal abnormalities in different phases of mitosis, seedling growth, and enzyme activity during germination are the preliminary investigations for plant bioassays. In this chapter, some of the most frequently and recently used methods for detection of genotoxicity with plant biosystems are reviewed.

2. Seed germination and root elongation tests

Many plant species have been recommended for ecotoxicity tests using seed germination and root elongation methods. Among them, cabbage, lettuce, and oats are recommended by the US Environmental Protection Agency (EPA) (1983) [18], the Organization for Economic Cooperation and Development (OECD) (1984) [19], and the Food and Drug Administration (FDA) (1987) [20]. Carrot, cucumber, and tomato are also suggested by the EPA and FDA, wheat is accepted by the FDA and OECD, and rice is also mentioned by the OECD. Although not mentioned in any of these documents, millet has been studied at the Illinois State Water Survey for several years [21]. Most frequently used species are *Allium cepa* L., *Lactuca sativa L.*, *Glycine max* (L.) Merr, *Avena sativa* L., *Hordeum vulgare* L., *Pisum sativum* L., *Tradescantia pallida* (Rose) D.R.Hunt , *Vicia faba* L., and *Zea mays* L. The crucifer *Arabidopsis thaliana* (L.) Heynh. is used only for mutation studies as its chromosomes are very small, and the total genome contains only about 70,000 kb in contrast to over a million kilobases in most other plants. The test substance, test duration, test organisms, the species and number of organisms, concentration of the test substance, replicates, randomization, equipment, reliability, environmental conditions (temperature, humidity, watering, lighting, photoperiod, and nutrients), observations, measurements, and analysis of the test results must be done carefully. The seed germination and seedling growth bioassays are more sensitive to separate plant developmental life stages as they integrate the effects of many environmental stress factors on both germination and seedling growth stages, respectively. The early seedling development is a more sensitive endpoint than the seed germination that depends on the energy reserves in cotyledons. Many researchers also found that the different kinds of species used do not respond similarly to toxic chemicals [22, 23]. Seed germination and plant growth bioassays are the most common techniques used to evaluate the toxicity of pesticides [24–27], heavy metals [6], allelochemicals [28], personal care products [29], compost [30], water samples taken from rivers [31], and industrial waste waters [25, 32]. Different plant species have also been used such as cucumber and cress [33], lettuce and soybean [34], red maple, sugar maple, white pine, and pink oak [35] for phytotoxicity tests.

3. Cytogenetic techniques

The frequency and the type of chromosome abnormalities in different phases of mitosis and the micronuclei frequency of interphase cells are analyzed by cytogenetic tests. The DNA damage caused by the genotoxic agents could either be repaired or otherwise could be lead to the DNA alterations. Chromosome abnormalities are the results of DNA double-strand breaks that were unrepaired or inaccurately repaired. Chromosomes are rearranged since broken chromosome ends become "sticky" and may combine with other broken chromosome ends. After mutagenic treatment, because of the chromosomal rearrangements and acentric fragments, dicentric bridges could be observed in mitotic cells of the first cell cycle. Micronuclei frequency also decreases in the interphase cell in the next cell cycle [36]. The micronucleus (MN) test, *A. cepa* and *V. faba* chromosome aberration test, and the T. MN tests have been recommended as the validated plant bioassays for laboratory testing and in situ monitoring of the genotoxicity of environmental mutagens [7]. Sister chromatid exchange (SCE) test can also be used to detect effects of small doses of pollutants; thus, it is adequate for initial genotoxicity evaluation tests [37]. SCEs result from alterations caused in the gene expression and by the loss of heterozygosity. SCE experiments are traditionally performed and well studied in mammalian cells. For plants, the protocols have been mainly developed in *V. faba* root cells [38].

3.1. *Allium/Vicia* chromosome aberration test

Several mutagens can be detected cytologically by cellular inhibition; disruption in metaphase; induction of chromosomal aberrations, numerical and structural, ranging from chromosomal fragmentation to the disorganization of the mitotic spindle; and consequently all subsequent dependent mitotic phases. The microscopic analysis includes mitotic index, micronuclei presence in interphase cells, and chromosomal aberrations in late anaphase and early telophase cells score. Approximately 1000 cells from all the stages of dividing cells in mitosis are counted in order to find the mitotic index value. Chromosomal abnormalities can be determined, and then, they are scored in the first 100 cells in different stages of mitotic division. The mostly used method to determine all of the abnormalities is to scan the slides from right to left, up, and down [39]. The *Allium* material is well known and has been used for the study of basic mechanisms as well as for scoring the effects of chemicals. *A. cepa* (the common onion) has proved to be the most useful and has repeatedly been suggested as a standard test material [40]. The use of *A. cepa* as a test system was introduced by Levan [41], when the effects of colchicine were investigated. Since then, the *Allium* test has been frequently used. Genotoxicity, cytotoxicity, and chromosome abnormalities in plant biosystems are mostly determined in *A. cepa* (2n = 16) and *V. faba* (2n = 12). They are efficient test organisms because of their availability throughout the year, ease of handling, and cultivation. They also do not need to be cultivated in sterile conditions; they have large and small number of chromosomes, which makes the observation of chromosomal damages in the mitotic cycle easier [42]. The *Allium* test has high sensitivity and good correlation when compared with the mammalian test systems. Ma and Grant [43] suggested including *Allium* test as a standard test system to determine chromosome damages induced by chemicals after the evaluation of 148 chemicals by the *Allium* test since 76% presented positive results. It was reported that the sensitivity of the *Allium* test was practically similar as the one observed for human lymphocyte and algae test systems. Rank and Nielsen [44] showed that the *Allium* test was more sensitive than the MicroScreen and the Ames tests. They also reported that there was a correlation of 82% of the carcinogenicity test in rodents in relation to the *Allium* test. The *V. faba* MN test has been shown to be sensitive in evaluating chromosomal aberrations and assessing genotoxicity from both organic and inorganic soil contaminants [45], sediment [46], organic material such as sewage sludge or composts [47] and water [48, 49]. Many researchers compared sensitivity of the *V. faba* test with other bioassays, i.e., somatic mutation and recombination test (SMART), that utilizes *D. melanogaster* Meigen. and compared with the *V. faba* sister chromatid exchange (SCE) test and MN inductions. Both tests showed 62.5% similarity [38]. Plant genotoxicity assays as the MN test on *V. faba* roots provide quantitative, repeatable, and reliable mutagenic data, and they are sensitive tests to detect new environmental mutagens or combination of different kinds of mutagens [50]. They can be used to develop new techniques for alternative assays in the determination of possible genetic damage caused by environmental pollutants such as pesticides, heavy metals, and more recently personal or health-care products. They can also contribute to an in situ monitoring, which can be carried out on a global scale in media as aqueous biota or soils in relation to human activities [1].

3.2. Tradescantia stamen hair mutation and micronucleus analysis

The genus *Tradescantia*, from the Commelinaceae family, is a higher plant with more than 500 species. Some of these and their clones are used as genetic bioindicators for mutagenic activity, such as *T. pallida* (Rose) D.R.Hunt, for environmental monitoring. It has two assay systems, the *Tradescantia* sp. staminal hair assay and the *Tradescantia* sp. MN assay, developed by Ma [51]. Stamen hair and MN tests have been widely employed for genotoxic effect studies with *Tradescantia* species [43, 52]. Almost all of the parts of the *Tradescantia* species including the root tip and also the pollen tube in development provide the best plant materials for cytogenetic toxicity testing studies. *Tradescantia* species have 12 chromosomes which are easily observable. Sax and Edmonds observed that meiotic chromosomes in pollen development were more easily influenced to breakage than mitotic chromosomes. They especially reported that the dividing chromosomes within the cells at meiosis are approximately ten times more sensitive to breakage than those in the interphase cells [42].

Ma and Grant [43] have prepared a historical perspective, detailing the importance of this plant in mutation studies. Firstly, the heterozygosity for flower color in *Tradescantia* sp. clones was used for these studies, and then, the stamen hairs have been determined to be good indicators of mutations. Clone 4430 is a hybrid of *Tradescantia hirsutiflora* Bush. and *Tradescantia subacaulis* Bush. reproduced only asexually, through cloning. This test uses the stamen hairs of *Tradescantia* sp. inflorescences to evaluate the frequency of somatic mutation, induced for mutagens, through changes in the color of stamen hair cells from blue to pink, due to the expression of a recessive gene of these cells. The frequency of micronuclei in tetrad cells of male meiotic cells in *Tradescantia* induced by the tested mutagen was determined [42]. The *Tradescantia* sp. MN test may be used for in situ exposure conditions to evaluate air or water pollution or under laboratory conditions for testing radioactive or chemical agents [53, 54]. The *Tradescantia* sp. stamen hair mutation (Trad-SH) assay (clone 4430) was evaluated for its efficiency and reliability as a screen for mutagens in an IPCS collaborative study on plant systems. The results of the study confirm that the Trad-SH assay is an unsuspicious system for screening potential environmental mutagens. A survey of the current literature indicates that the Trad-SH assay could be used for in situ monitor of liquid, gaseous, and also radioactive pollutants as well although the study was carried out under laboratory conditions [55].

3.3. Sister chromatid exchange

The sister chromatid exchange (SCE) test is developed from the semiconservative DNA replication model which we could see the separation of DNA. The cytogenetic monitoring of exposure to potential mutagens in the environment could be done by SCE which is a highly sensitive cytogenetic tool for detecting DNA damage. It involves firstly the breakage of both DNA strand and then an exchange of whole DNA duplexes. The symmetrical exchange during S phase at one locus between sister chromatids that does not alter chromosome length and genetic information is defined. Taylor was the first scientist who made the SCE test visualized for plant cells, but he used tritium and autoradiography, which provided poor spatial resolution [56]. After Taylor, it was discovered that sister chromatids could be differentiated and revealed SCEs in combination with Hoechst dye 33258 incorporation of the DNA base analog 5′-bromodeoxyuridine (BrdUrd) staining [57]. BrdUrd is a synthetic nucleoside that is an analog of thymidine and is actively incorporated into the newly synthesized DNA during replication process. It is commonly used in the detection of dividing cells in living organisms during the S phase of the cell cycle substituting for thymidine. The standard fluorescence plus Giemsa (FPG) staining method also will enable visualization of SCEs in metaphase spreads of growing cells in medium containing BrdUrd with a light microscope [56]. The frequency of SCEs per chromosome set increases after treatment with genotoxic agents. SCE method was first applied in mammalian cells, and later, it has been shown that it can be applied in plant cells.

Especially plant species that have relatively large and a low number of chromosomes such as *A. cepa* and *V. faba* are used for SCE analysis [57, 58]. *Crepis capillaris* (L.) Wallr. is also a good material for analyzing the frequency of SCE with 2n = 6 chromosome number. It allows studying SCE frequency in each chromosome type, since it has three pairs of morphologically differentiated chromosomes [59, 60].

4. Molecular techniques

4.1. Fluorescent in situ hybridization

The classical cytogenetic techniques were usually used for detecting the changes in chromosomal number and morphology. However, chromosome staining with the traditional methods such as Feulgen or orcein staining can fail in the analysis of small changes in chromosome structure. The fluorescent in situ hybridization (FISH) allows the detection and a more detailed localization of chromosomal rearrangements, both in interphase and mitotic nuclei, which gives new possibilities to study chromosomal aberrations [61]. Additionally, it helps to reveal the mechanisms of the formation of chromosomal abnormalities in plant mutagenesis. Although there are a few number of DNA probes for particular plant chromosomes, *A. thaliana* is a good example when FISH employing chromosome region-specific DNA probes (e.g., centromere, telomere, rDNA) is helpful in chromosome aberration analysis. The translocations in chromosomes of tetraploid plants of *A. thaliana* have been detected by FISH [62]. The effects of maleic acid hydrazide on hairy root tip meristem cells of *C. capillaris* were studied with FISH using rDNA and telomeric sequences as a probe and spontaneous chromosomal rearrangements were determined [63]. It is also important to analyze the chromosomal rearrangements in interphase cells treated with mutagenic chemicals that may cause a decrease in the frequency of cell divisions. The basic steps of this procedure are the same as the other organisms, but several cytogenetic laboratories modified various techniques for plant cells.

4.2. TUNEL test

Another test used to identify apoptosis that has found application in plant genotoxicity studies is the terminal deoxyribonucleotidyl transferase-mediated deoxyuridine triphosphate nick end labeling (TUNEL) test [64]. TUNEL assay detects DNA fragmentation by the help of fluorescence microscope. TUNEL test is used to detect DNA damage associated with nonapoptotic events such as necrotic cell death induced by exposure to genotoxic chemicals. It is not limited to the detection of apoptotic cells [65] and has also ability to stain cells going through active DNA repair [66]. The regulated cell death plays an important role during development of plants, and it is also essential for plant-specific responses to biotic and abiotic stress factors. The terminal deoxynucleotidyl transferase catalyzes the polymerization of labeled nucleotides to DNA strand breaks in situ. For TUNEL test, successive hand-cut sections of each axis of embedded plant material are stained with propidium iodide (PI) in order to stain the nuclei of dead cells to red and DAPI (4',6-diamidino-2-phenylindol) which can pass through the normal cell membrane and stains the nuclei to blue. DAPI can be used to stain both live and fixed cells. The detection of DNA breaks at a single nucleus can be achieved with TUNEL test within a short time, and the screening of labeled nuclei is easier than other methods. It is recommended for the preliminary genotoxicity investigation of the new identified chemicals [67].

4.3. Single-cell gel electrophoresis (comet assay)

DNA damage in higher plant cells was evaluated by the frequency of chromosomal aberrations in metaphase chromosomes, abnormal anaphase and telophases, and micronuclei; however, these tests measure unrepaired genome damage in cells which have reached mitosis. DNA damage may be originated from DNA metabolism spontaneously or from the effects of environmental factors. There are different kinds and levels of DNA repair mechanisms in cell nucleus to prevent these damages. When the repair mechanisms are ineffective or there was a heavy DNA damage, it may lead to the inhibition of replication, transcription, or protein synthesis; however, in the long term, chromosomal abnormalities or mutations could be formed. It is a sensitive and fast fluorescent technique, which is used to determine the amount of DNA damage on single cell level. After its introduction as "alkaline comet assay," it has been developed with many modifications for investigating the process of apoptosis and became a workable technique for detecting a variety of DNA damages in plant cells. It allows the determination of double- and single-stranded DNA breaks in a single cell and also helps to measure the level of the migration of DNA by using horizontal gel electrophoresis system [68]. The length of the tail and the amount of the DNA in the head and in the tail are measured to assess the toxicity in a computerized image analysis system. The tail moment (TM) can be calculated to show DNA damage [69]. The comet assay allows fast detection of DNA damage, shortly after the injury, before DNA is repaired, and without any need to wait for progression into mitosis [70]. The presence of a cell wall and the absence of free cells in plant tissues cause technical difficulties for performing the comet assay. Over the past few years, many scientists have improved the methodology for the comet assay on plant cells. Navarrete et al. [70]. developed a simple and efficient mechanical extraction to isolate cell nuclei to overcome these problems. This technique was then improved by Gichner. The different internal parameters such as nucleus isolation methods, filtration and lysis steps, agarose concentration, and the external parameters such as room temperature and light intensity were evaluated during these studies [71].

4.4. Random amplified polymorphic DNA polymerase chain reaction (RAPD-PCR) technique

RAPD-PCR is a PCR-based and quite reproducible technique that yields information on a large number of markers without having to obtain DNA sequence information for primer design [72]. Many scientists used RAPD-PCR technique commonly for a variety of purposes such as cultivar identification, genetic diversity assessment, and the construction of phylogenetic relationships [73], and it has been successfully utilized in genotoxicity identification of toxic chemicals. A number of selective and sensitive assays for DNA analysis in ecotoxicology have been developed with the improvement of recent molecular biology techniques. DNA-based techniques such as RFLP, QTL, RAPD, AFLP, SSR, and VNTR are being used to investigate the variations at the DNA sequence level. RAPD-PCR can be used to detect genotoxicity, and differences in RAPD profiles can clearly be shown when comparing DNA fingerprints from untreated and treated individuals to genotoxic agents [74]. Many studies support the view that the RAPD analysis is a highly sensitive method for the detection of DNA damage induced by environmental pollutants like toxic chemicals. RAPD markers are at this moment low valuable markers due to the lack of repeatability. A few work is usually published at this moment using this kind of markers. This kind of study using other DNA markers will be of much more interest.

4.5. Real-time polymerase chain reaction (RT-PCR) technique

Plants have risk of DNA damage due to continuous exposure to environmental mutagens, and thus a variety of repair mechanisms should operate to maintain genome integrity. *A. thaliana* is a mostly studied plant for the repair mechanisms after exposure to several mutagens such as UV-B radiation [75], heavy metal contamination [76], and wound stress [77]. In the first step of the DNA damage response, DNA lesions or replication inhibition must be detected. The DNA damage response is controlled by the activation of several regulatory kinases and also checkpoint proteins that lead to specific cell cycle arrests as well as changes in the chromatin structure at the site of DNA damage. The transcriptional regulation of the genes could be determined by RT-PCR in order to evaluate the mechanism of plant response to genotoxic agents. To investigate effects of mutagens on the transcript levels of some gene-encoding antioxidative enzymes, such as superoxide dismutase (SOD), catalase (CAT), and ascorbate peroxidase (APX), they were studied in Ref. [78]. The mutants of *A. thaliana* that are hypersensitive to UV radiation (designated uvh and uvr) have been isolated to investigate the respond of plants and its pathways to UV radiation. UVR2 and UVR3 products were previously identified as photolyases that remove UV-induced pyrimidine dimers in the presence of visible light [76]. Hu et al. (2007)

investigated the role of calmodulin (CaM) and the relationship between CaM and hydrogen peroxide (H_2O_2) in abscisic acid (ABA)–induced antioxidant defense in leaves of *Z. mays* [78].

5. Flow cytometry

Flow cytometry (FCM) is a rapid and multiparametric technique that theoretically has the potential to detect minute variations in nuclear DNA (nDNA) content, as well as chromosomal damage, in exposed organisms. It can also provide information on polyploidization and evaluate cell cycle dynamics in plants. Pfosser et al. [79]. evaluate the sensitivity of FCM by detecting the variations in DNA content as small as 1% in aneuploid wheat-rye lines. Relatively to DNA damage, Rayburn and Wetzel correlated the coefficient of variation of the G0/G1 peak with chromosomal aberration in aluminum-exposed plants, as this parameter is able to detect broken and rearranged chromosomes in daughter cells [80]. Monteiro et al. also detected an increase in the full peak coefficient of variation (FPCV) of the G0/G1 peak of lettuce plants exposed to Cd [81].

6. Conclusion

Hundreds of new industrial chemicals have been continuously produced to facilitate our lives, and we are not able to be aware of their damage before we investigate their effects on organisms. Plant bioassays serve as a tool to demonstrate the cytotoxic and genotoxic effects of environmental pollutants by means of clear-cut evidence of chromosome damage and gene mutation. These studies could also be useful to establish a database for environmental conditions in the various regions of the world. Some of these simple and clear-cut indicators revealed by plant bioassays could also be used to demonstrate the genotoxic effects of environmental pollution to the general public. The kind of education that is required is not only about teaching people how to detect and eliminate pollutants but also to educate the general public on the root cause of pollution problems. Pollution is related to every facet of human life, and it is life itself that generates pollution. Regulations and guidelines are essential to cure the symptoms of pollution. Plant bioassay studies deserve to be included by the enforcement agencies, particularly of the developing countries, for their regular monitoring of pollution sites.

Author details

Özlem Aksoy

Address all correspondence to: ozlem.aksoy@kocaeli.edu.tr, odalgic@yahoo.com

Department of Biology, Faculty of Science and Literature, Kocaeli University, İzmit, Kocaeli, Turkey

References

[1] Grant WF. The present status of higher plant bioassays for the detection of environmental mutagens. Mutation Research-Fundamental and Molecular Mechanisms of Mutagenesis. 1994;**310**(2):175-185. DOI: 10.1016/0027-5107(94)90112-0

[2] Büyük İ, Soydam-Aydın S, Aras S. Molecular responses of plants to stress conditions. Türk Hijyen ve Deneysel Biyoloji Dergisi. 2012;**69**(2):97-110. DOI: 10.5505/TurkHijyen.2012.40316

[3] Constantin MJ, Owens ET. Introduction and perspective of plant genetic and cytogenetic assays. Mutation Research. 1982;**99**:37-49

[4] Grant WF, Salamone MF. Comparative mutagenicity of chemicals selected for test in the International Program on Chemical Safety`s collaborative study on plant systems for the detection of environmental mutagens. Mutation Research. 1994;**310**:187-209

[5] Ma TH. The international program on plant bioassays collaborative studies with plant systems. Mutation Research. 1999;**426**:97-98

[6] Knasmuller S, Gottmann E, Steinkellner H, Fomin A, Pickl C, Paschke A, God R, Kundi M. Detection of genotoxic effects of heavy metal contaminated soils with plant bioassays. Mutation Research. 1998;**420**:37-48

[7] Gopalan HNB. Ecosystem health and human wellbeing: the mission of the International Program on Plant Bioassays. Mutation Research. 1999;**426**:99-102

[8] Sandhu SS, de Serres FJ, Gopalan HNB, Grant WF, Velemisky J, Becking GC. An introduction and study design. Mutation Research. 1994;**10**:169-173

[9] Liu Y, Liu J, Zhang A, Liu Z. Treatment effects and genotoxicity relevance of the toxic organic pollutants in semi-coking wastewater by combined treatment process. Environmental Pollution. 2017;**220**(Pt A);13-19. DOI: 10.1016/j.envpol.2016.04.095

[10] Fadic X, Placencia F, Domínguez AM, Cereceda-Balic F. Tradescantia as a biomonitor for pesticide genotoxicity evaluation of iprodione, carbaryl, dimethoate and 4,4′-DDE. The Science of the Total Environment. 2017;**575**:146-151. DOI: 10.1016/j.scitotenv.2016.09.198

[11] Tao N, Liu G, Bai L, Tang L, Guo C. Genotoxicity and growth inhibition effects of aniline on wheat. Chemosphere. 2017;**169**:467-473. DOI: 10.1016/j.chemosphere.2016.11.063

[12] Quintana VR, Gómez-Arroyo S, Sánchez-Alarcón J, Milić M, Olivares JL, Waliszewski SM, Cortés-Eslava J, Villalobos-Pietrini R, Calderón-Segura ME. Genotoxic effects of the carbamate insecticide Pirimor-50® in Vicia faba root tip meristems and human lymphocyte culture after direct application and treatment with its metabolic extracts. Arhiv za Higijenu Rada i Toksikologiju. 2016:**67**(4):266-276. DOI: 10.1515/aiht-2016-67-2809

[13] Zhen C, Moshun C, Ming J. Hydrogen sulfide alleviates mercury toxicity by sequestering it in roots or regulating reactive oxygen species productions in rice seedlings. Plant Physiology and Biochemistry. 2017;**111**:179-192

[14] Scaldaferro MA, Prina AR, Moscone EA, Kwasniewska J. Effects of ionizing radiation on *Capsicum baccatum* var. pendulum (Solanaceae). Applied Radiation and Isotopes. 2013;**79**:103-108. DOI: 10.1016/j.apradiso.2013.04.004

[15] Na roi-et V, Chiemchaisri W & Chiemchaisri C. Genotoxicity assessment of volatile organic compounds in landfill gas emission using comet assay in higher terrestrial plant. Bulletin of Environmental Contamination and Toxicology. 2016;98:283. DOI: 10.1007/s00128-016-2012-3

[16] Venkatachalam P, Jayalakshmi N, Geetha N, Sahi SV, Sharma NC, Rene ER, Sarkar SK, Favas PJC. Accumulation efficiency, genotoxicity and antioxidant defense mechanisms in medicinal plant Acalypha indica L. under lead stress. Chemosphere. 2017;**171**: 544-553

[17] Uhl M, Plewa M, Majer BJ, Knasmuller S. Basic principles of genetic toxicology with an emphasis on plant bioassays. In: Maluszynska J, Plewa M, editors. Bioassays in Plant Cells for Improvement of Ecosystem and Human Health. Katowice: Wydawnictwo Uniwersytetu Śląskiego; 2003. pp. 11-30

[18] Ratsch HC. Interlaboratory Root Elongation Testing of Toxic Substances on Selected Plant Species. EPA-600/S3-83-051. Corvallis, OR: U.S. Environmental protection Agency; 1983

[19] Organization for Economic Cooperation and Development. Terrestrial Plants: Growth test. Paris: OECD guideline for Testing Chemicals; 1984. No: 208

[20] Food and Drug Administration. Seed Germination and Root Elongation. Environmental Assessment Technical Assistance Document 4.06, the Center for Food Safety and Applied Nutrition and the Center for Veterinary Medicine. Washington, D.C.: U.S. Department of Health and Human Services; 1987

[21] Wang W, Keturi PH. Comparative seed germination tests using ten plant species for toxicity assessment of a metal engraving effluent sample. Water, Air, & Soil Pollution. 1990;**52**:369-376. DOI: 10.1007/BF00229444

[22] Pfleeger T, Mc Farlane C, Sherman R, Volk G. A short-term bioassay for whole plant toxicity. In: Gorsuch JW, Lower WR, Wang W, Lewis MA. editors. Plants for Toxicity Assessment. Vol. 2 (ASTM STP 1115). Philadelphia, PA: American Society for Testing and Materials; 1991. pp. 355-364

[23] Gong P, Wilke BM, Fleischmann S. Soil-based phytotoxicity of 2,4,6-trinitrotoluene to higher plants. Archives of Environmental Contamination and Toxicology. 1999;**36**: 152-157

[24] Dane F, Dalgıç Ö. The effects of fungicide benomyl (Benlate) on growth and mitosis in onion (*Allium cepa* L.) root apical meristem. Acta Biologica Hungarica. 2005;**56**:119-128

[25] Aksoy Ö, Erbulucu T, Vatan E. Effects of wastewater from olive oil and milk industry on growth and mitosis in *Allium cepa* root apical meristem. Journal of Applied Biological Sciences. 2011;**5**(3):75-78

[26] Aksoy Ö, Deveci A. The investigation of the cytotoxic effects of some pesticides on soybean (*Glycine max* L.). Cytologia. 2012;**77**(4):475-483

[27] Tütünoğlu B, Aksoy Ö. Morphological and anatomical effects of the herbicide roundup on *Zea mays* (Poaceae). Journal of Functional and Environmental Botany. 2014;**4**(2):71-76

[28] Zackrisson O, Nilsson MC. Allelopathic effects by *Empetrum hermaphroditum* on seed germination of two boreal tree species. Canadian Journal of Forest Research. 1992;**22**(9):1310-1319. DOI: 10.1139/x92-174

[29] An J, Zhou Q, Sun Y, Xu Z. Ecotoxicological effects of typical personal care products on seed germination and seedling development of wheat (*Triticum aestivum* L.). Chemosphere. 2009;**76**:1428-1434

[30] Kapanen A, Itavaara M. Ecotoxicity tests for compost applications. Ecotoxicology and Environmental Safety. 2001;**49**:1-16

[31] Aksoy Ö, Erbulucu T, Öner S, Tekeli İB. Phytotoxic And genotoxic effects of water samples taken from the eastern channel of Kocaeli on *Vicia faba* and *Zea mays*. Fresenius Environmental Bulletin. 2012;**21**(7):1820-1826

[32] Aksoy Ö. The phytotoxic and genotoxic effects of olive oil and milk industry waste water on *Elodea canadensis*. Journal of Functional and Environmental Botany. 2013;**3**(1): 43-53

[33] Helfrich P, Chefetz B, Hadar Y, Chen Y, Schnabl H. A novel method for determining phytotoxicity in compost. Compost Science and Utilization. 1998;**6**:6-13

[34] Gundersson CA, Kostuk JM, Mitcell HG, Napolitano GE, Wicker LF, Richmond JE, Stewart AJ. Multispecies toxicity assessment of compost produced in bioremediation of an explosives-contaminated sediment. Environmental Toxicology and Chemistry. 1997;**16**:2529-2537

[35] Maynard AA. Utilization of MSW compost in nursery stock production. Compost Science and Utilization. 1998;**6**:38-44

[36] Maluszynska J, Juchimiuk J. Plant genotoxicity: A molecular cytogenetic approach in plant bioassays. Arhiv za Higijenu Rada i Toksikologiju. 2005;**56**:177-184

[37] Eastmond DA. Sister chromatid exchanges. In: Wexler P. editor. Encyclopedia of Toxicology. 2nd ed. San Diego: Academic Press. 2005. pp. 19-20

[38] Iqbal M. *Vicia faba* bioassay for environmental toxicity monitoring: A review. Chemosphere. 2016;**144**:785-802

[39] Feretti D, Zerbini I, Zani C, Ceretti E, Moretti M, Monarca S. *Allium cepa* chromosome aberration and micronucleus tests applied to study genotoxicity of extracts from pesticide-treated vegetables and grapes. Food Additives and Contaminants. 2007;**24**(6):561-572

[40] Stich HF, Lam P, Lo LW, Koropatnick DJ and San HC. The search for relevant short-term bioassays for chemical carcinogens: The tribulation of a modern sisyphus. Canadian Journal of Genetics and Cytology. 1975;**17**:471-492

[41] Levan A. The effect of colchicine on root mitoses in Allium. Hereditas. 1938;**24**:471-486

[42] De Souza CP, Guedes TA, Fontanetti CS. Evaluation of herbicides action on plant bioindicators by genetic biomarkers: a review. Environmental Monitoring and Assessment. 2016;**188**(12):694. DOI: 10.1007/s10661-016-5702-8

[43] Ma TH, Grant WF. The Tradescantias—adventurous plants. The Herbarist. 1982;**48**:36-44

[44] Rank J, Nielsen MH. Evaluation of the Allium anaphase–telophase test in relation to genotoxicity screening of industrial wastewater. Mutation Research. 1994;**312**:17-24

[45] Cotelle S, Masfaraud JF, Férard JF. Assessment of the genotoxicity of contaminated soil with the Allium/Vicia-micronucleus and the Tradescantia-micronucleus assays. Mutation Research. 1999;**426**:167-171

[46] Chen G, White PA. The mutagenic hazards of aquatic sediments: A review. Mutation Research. 2004;**567**:151-225

[47] De Simone C, Tomati U, Galli E, Owczarek M, De Marco A, D'Ambrosio C, Alianello F, Cortellini L. Evaluation of toxic and genotoxic activity of some composts from different origin. Fresenius Environmental Bulletin. 2009;**9**:683-690

[48] Miao CM, Fu R, Yang D, Zheng L. Vicia root micronucleus assay on the clastogenicity of water samples from the Xiaoqing River in Shandong Province of the People's Republic of China. Mutation Research. 1999;**426**:143-145

[49] Monarca S, Zani C, Richardson SD, Thruston AD, Moretti JR, Feretti MD, Villarini MA. New approach to evaluating the toxicity and genotoxicity of disinfected drinking water. Water Reserach. 2004;**38**:3809-3819

[50] Khadra A, Pinelli E, Lacroix MZ, Bousquet-Melou A, Hamdi H, Merlina G, Guiresse M, Hafidi M. Assessment of the genotoxicity of quinolone and fluoroquinolones contaminated soil with the *Vicia faba* micronucleus test. Ecotoxicology and Environmental Safety. 2012;**76**(2):187

[51] Ma TH. The role of plant systems for the detection of environmental mutagens and carcinogens. Mutation Research. 1999;**437**:97-100

[52] Rodrigues GS. Bioensaios de toxicidade genética com Tradescantia. Jaguariúna: Embrapa Meio Ambiente; 1999

[53] Gichner T, Veleminský J. Monitoring the genotoxicity of soil extracts from two heavily polluted sites in Prague using Tradescantia stamen hair and micronucleus (MNC) assays. Mutation Research. 1999;**426**:163-166

[54] Cebulska–Wasilewska A, Plewa MJ. Short term bioassay based on gene mutations level in Tradescantia cells (Trad-SH). In: Maluszynska J, Plewa M, editors. Bioassays in plant cells for improvement of ecosystem and human health. Katowice: Wydawnictwo Uniwersytetu Śląskiego. 2003. pp: 31-38

[55] Ma TH, Cabrera GL, Gill BS, Sandhu SS, Vandenberg AL, Salamone MF. Tradescantia micronucleus bioassay. Mutation Research. 1994;**310**:221-230

[56] Wilson DM, Thompson LH. Molecular mechanisms of sister-chromatid exchange. Mutation Research. 2007;**616**:11-23 DOI: 10.1016/j.mrfmmm.2006.11.017

[57] Cortès F, Andersson HC. Analysis of SCEs in *Vicia faba* chromosomes by a simple fluorescent plus Giemsa technique. Hereditas. 1987;**107**:7-13

[58] Cortès F, Escalza P, Mateos S, Diaz-Recasens M. Factors affecting the production of SCEs by maleic hydrazide in root-tip chromosomes of *Allium cepa*. Mutation Research. 1987;**192**:125-130

[59] Dimitrov B. Relationship between sister-chromatid exchanges and heterochromatin or DNA replication in chromosomes of Crepis capillaris. Mutation Research. 1987;**190**:271-276

[60] Maluszynska JB. Chromosomes of *Crepis capillaris* (L.) Waller. in vivo and in vitro. Katowice: Prace Naukowe Uniwersytetu Śląskiego nr 147; 1990

[61] Kolano B, Wolny E, Maluszynska J. Fluorescent in situ hybridisation in plant mutagenesis. In: Maluszynska J, Plewa M, editors. Bioassays in plant cells for improvement of ecosystem and human health. Katowice: Wydawnictwo Uniwersytetu Śląskiego; 2003. pp: 139-148

[62] Weiss H, Maluszynska J. Chromosomal rearrangement in autotetraploid plants of *Arabidopsis thaliana*. Hereditas. 2000;**133**:255-261

[63] Maluszynska J, Juchimiuk J, Wolny E. Chromosomal aberrations in *Crepis capillaris* cell selected by FISH. Folia Histochem et Cytobiologia. 2003;**41**:101-104

[64] Havel L, Durzan DJ. Apoptosis in plants. Botanica Acta: Journal of the German Botanical Society. 1996;**109**:268-277

[65] Ansari B, Coates PJ, Greenstein BD, Hall PA. In situ end labeling detects DNA strand breaks in apoptosis and other physiological and pathological states. Journal of Pathology. 1993;**170**:1-8

[66] Kanoh M, Takemura G, Misao J, Hayakawa Y, Aoyama T, Nishigaki K, Noda T, Fujiwara T, Fukuda K, Minatoguchi S, Fujiwara H. Significance of myocytes with positive DNA in situ nick end-labeling (TUNEL) in hearts with dilated cardiomyopathy. Not apoptosis but DNA repair. Circulation. 1999;**99**:2757-2764

[67] Juchimiuk J, Maluszynska J. Detection of DNA fragmentation caused by chemical mutagens using the TUNEL test. In: Maluszynska J, Plewa M, editors. Bioassays in Plant Cells for Improvement of Ecosystem and Human Health. Katowice: Wydawnictwo Uniwersytetu Śląskiego; 2003. pp. 133-138

[68] Jaloszynski P, Kujawski M, Szyfter K. Elektroforeza pojedynczych komórek (comet assay) – użyteczna technika badania uszkodzeń DNA. Postępy Biologii Komórki. 1996;**23**:339-354

[69] Gichner T. Comet assay in higher plants. In: Maluszynska J, Plewa M, editors. Bioassays in plant cells for improvement of ecosystem and human health. Katowice: Wydawnictwo Uniwersytetu Śląskiego; 2003. pp. 123-132

[70] Navarrete MH, Carrera P, Miguel M, Torre C. A fast comet assay variant for solid tissue cells. The assessment of DNA damage in higher plants. Mutation Research/Genetic Toxicology and Environmental Mutagenesis. 1997;**389**:271-277

[71] Pourrut B, Pinelli E, Celiz Mendiola V, Silvestre J, Douay F. Recommendations for increasing alkaline comet assay reliability in plants. Mutagenesis. 2015;**30**(1):37-43. DOI: 10.1093/mutage/geu075

[72] Agarwal M, Shrivastava N, Padh H. Advances in molecular marker techniques and their applications in plant sciences. Plant Cell Reports. 2008;**27**:617-631

[73] Aksoy Ö, Erbulucu T, Özen F, Deveci A. Genetic variation in critically endangered plant *Amsonia orientalis* Decne. Journal of Biodiversity and Environmental Sciences. 2013;**3**(1):44-53

[74] Cenkci S, Yıldız M, Ciğerci İH, Konuk M, Bozdağ A. Toxic chemicals-induced genotoxicity detected by random amplified polymorphic DNA (RAPD) in bean (*Phaseolus vulgaris* L.) seedlings. Chemosphere. 2009;**76**:900-906

[75] Liu W, Yang YS, Zhou Q, Xie L, Li P, Sun T. Impact assessment of cadmium contamination on rice (*Oryza sativa* L.) seedlings at molecular and population levels using multiple biomarkers. Chemosphere. 2007;**67**:1155-1163

[76] Liu Z, Hossain GS, Islas-Osuna MA, Mitchell DL, Mount DW. Repair of UV damage in plants by nucleotide excision repair: Arabidopsis UVH1 DNA repair gene is a homolog of *Saccharomyces cerevisiae* Rad1. The Plant Journal. 2000;**21**:519-528. DOI: 10.1046/j.1365-313x.2000.00707

[77] Mediouni C, Houlné G, Chaboute ME, Ghorbel MH, Jemal F. Cadmium and copper genotoxicity in plants. In: Biosaline Agriculture and High Salinity Tolerance. Basel: Birkhäuser; 2008. pp. 325-333. DOI: 10.1007/978-3-7643-8554-5-30

[78] Hu X, Jiang M, Zhang J, Zhang A, Lin F, Tan M. Calcium–calmodulin is required for abscisic acid-induced antioxidant defense and functions both upstream and downstream of H_2O_2 production in leaves of maize (*Zea mays*) plants. New Phytologist. 2007;**173**:27-38. DOI: 10.1111/j.1469-8137.2006.01888

[79] Pfosser M, Heberle-Bors E, Amon A and Lelley T. Evaluation of sensitivity of flow cytometry in detecting aneuploidy in wheat using disomic and ditelosomic wheat–rye addition lines. Cytometry. 1995;**21**:387-393. DOI:10.1002/cyto.990210412

[80] Lane Rayburn A, Wetzel JB. Flow cytometric analyses of intraplant nuclear DNA content variation induced by sticky chromosomes. Cytometry. 2002;**49**(1):36-41. DOI:10.1002/cyto.10135

[81] Monteirono MS, Rodriguez E, Loureiro J, Mann RM, Soares AMVM & Santos C. Flow cytometric assessment of Cd genotoxicity in three plants with different metal accumulation and detoxification capacities. Ecotoxicology and Environmental Safety. 2010;**73**:1231-1237

Chapter 3

Ecology of Woody Plants in African Savanna Ecosystems

Allan Sebata

Additional information is available at the end of the chapter

http://dx.doi.org/10.5772/intechopen.69865

Abstract

Woody plants are key components of African savanna ecosystems as they provide wildlife habitats, offer browsing to ungulates and are also a major source of fuel wood. Disturbance events such as herbivory and fire negatively affect woody plant communities. However, some woody plants respond to disturbance events through resprouting. In savanna ecosystems, woody plants co-occur with grasses and disturbance events such as overgrazing result in the proliferation of woody plants at the expense of the grasses. Therefore, an understanding of the factors that influence woody plants is critical for the better management of African savanna ecosystems. This chapter reviewed our current knowledge of the ecology of woody plants in African savanna ecosystems and examined disturbance events such as herbivory and fire that shape woody plant communities. The role of resprouting as a response to disturbance events and the negative effects of woody plant encroachment on African savannas was also investigated. In addition, the consequences of poor management such as woody plants loss and possible restoration measures were explored. Disturbance events such as herbivory and fire were found to play critical roles in shaping the African savanna ecosystems. Interventions such as restoration have a role to play in restoring the productivity of degraded woody plant communities.

Keywords: encroachment, fire, herbivory, resprouting, restoration, savanna

1. Introduction

In most African savannas, plant communities are influenced by shortage of moisture during the dry season, with growth occurring largely during the wet season. The occurrence of dry seasons and the resultant fires, fuelled by a continuous annual supply of dry fuel, are thought to be the key drivers in the development of African savannas. Savannas occur where rainfall is seasonal and unpredictable. In general, savannas are characterised by a continuous grass layer which is occasionally interrupted by woody plants, with fires occurring from time to time [1, 2]. Furthermore, plant communities in the savanna evolved under and continue to be increasingly subjected to intense herbivory pressures. African savanna ecosystems are an important wildlife habitat, offer grazing to livestock and are also a major resource for fuel wood and other products. Their structure and productivity are determined by complex and dynamic interactions between climate, soils and disturbances (such as fire and herbivory) [3]. Woody plants in savannas create favourable micro environments (e.g. through deposition of leaf litter and shading) and habitats that can support a great diversity of flora and fauna [4]. Woody plant communities in African savannas are influenced by many factors such as rainfall, soil type, herbivory and fire. The ability of woody plant communities to cope with disturbances is critical for the sustainability of African savanna ecosystems. Resprouting is widely acknowledged as a mechanism through which woody plants respond to disturbance events such as fire and herbivory. The productivity of African savanna ecosystems is negatively affected by the proliferation of woody plants, a phenomenon referred to as woody plant encroachment. An understanding of the factors that favour woody plant encroachment is important for the better management of African savannas. Poor management of woody plant communities in African savannas leads to land degradation with restoration a slow and expensive process.

This chapter was based on a review of the current literature and sought to highlight the state of our knowledge on the ecology of woody plants in African savanna ecosystems. An extensive search for literature on the effects of rainfall, soil type, herbivory and fire on woody plant communities was undertaken. The role of resprouting as a mechanism that enables woody plants to cope with disturbance events such as fire and herbivory was also examined. Additionally, the negative effects of woody plant encroachment on African savanna communities were covered. Finally, land degradation as a consequence of poor woody plant community management together with possible restoration measures was discussed.

2. Effect of rainfall on woody plant communities in African savannas

Woody vegetation structure is determined by the amount of precipitation, with many African savannas water-limited [5–7]. Gordijn et al. [8] reported an increase in woody vegetation with increasing mean annual rainfall. Savannas can be classified according to the amount of rainfall, length of dry season and reliability of rainfall as shown in **Table 1**.

Types of savanna	Mean annual rainfall (mm)	Length of dry season (months)	Normal deviation of rainfall from the annual mean (%)
(a) Moist savanna	1000–2000	2.5–5	15–20
(b) Dry savanna	500–1000	5–7.5	20–25
(c) Semi-arid savanna	250–500	7.5–10	25–40

Table 1. Classification of savannas according to amount of rainfall received per year, length of dry season and reliability of rainfall.

Rainfall is least predictable in the semi-arid savanna and most predictable in the moist savanna. Water is considered the main resource limiting plant growth in semi-arid savannas [9]. Woody plant biomass, basal cover and height increase with increasing availability of water to the plants [10]. For example, woody plants are usually abundant along drainage lines within semi-arid savannas owing to greater availability of water [11]. There is also greater woody plant species richness and equitability with increasing rainfall [10, 12]. In addition to the amount of moisture available to plants, the spatio-temporal distribution of water will determine the actual species present and how they are distributed in space. The effects of rainfall interact with soil nutrients, fire and herbivory to influence the density of woody plants [13, 14]. In savannas, the extent of woody vegetation cover at a regional scale is determined by precipitation, while at the landscape level it is influenced by geologic substrate, topography, fire and large herbivores, especially elephants. The density of woody plants varies from dense in woodlands to sparse in nearly treeless areas [15]. Woody plant cover is a key determinant of ecosystem function in savannas [6]. Sankaran et al. [6] set a threshold of 650 mm mean annual precipitation as limiting woody plant growth, above which maximum closed woody cover canopy can be achieved. Additionally, the stature of woody plants decreases with declining precipitation to the point where below *ca.* 300 mm most woody plants will be shorter than the arbitrary 2.5 m threshold used to distinguish trees from shrubs. Scholes et al. [10] found that members of the *Mimosaceae* (mainly *Acacia*) to dominate the tree layer in areas with mean annual precipitation (MAP) of up to 400 mm were then replaced by either the *Combretaceae* (*Combretum* or *Terminalia*) or *Colophospermum mopane* of the *Caesalpiniaceae* where MAP was between 400 and 600 mm and by other representatives of the *Caesalpiniaceae* above 600 mm MAP. Although high precipitation results in increased recruitment of woody plants [16], other factors such as fire, herbivory and frost preclude woody vegetation from reaching the maximum woody cover [17]. A combination of frost, fire and herbivory (for example by elephants) are important determinants of the structure and composition of the woody vegetation of some southern African savannas [18, 19].

3. Effect of soil type on woody plant communities in African savannas

The spatial heterogeneity of woody vegetation in African savannas is influenced by the physical and chemical properties of soil [20]. For instance, shallow, gravelly soils with a low soil nutrient status will limit the woody plant size. Soil moisture and nutrient content are related to geology [21], implying that geology predetermines the array of vegetation types found in the African savannas [22]. For example, in African savannas, broad-leafed savanna occur on ancient, highly weathered surfaces, whereas the fine-leafed savanna is restricted to recently formed, nutrient-rich soils [14, 23]. The *Combretaceae* (*Terminalia & Combretum*) make up about half of the basal area on soils that are free-draining or rocky, whereas soils with an impeded layer (often sodic or calcareous) within the rooting zone are dominated by *Colophospermum mopane* (Kirk ex Benth.) Kirk ex J. Leonard, an ecologically and morphologically atypical member of *Caesalpiniaceae* [10]. Furthermore, sandy soils tend to favour woody over herbaceous (grasses) plants, which could be attributed to their ability to allow water to percolate deeper beyond the rooting zone of grasses [6, 14, 24, 25]. Additionally, woody plant cover declines as soil clay content increases [26], because the higher water holding capacity of the finer textured clay soils favours the shallow-rooted grasses over the deep-rooted woody vegetation [17]. Scholes [27] also reported nutrient-poor savannas as generally supporting higher woody biomass than nutrient-rich ones.

4. Effect of herbivory on woody plant communities in African savannas

In African savanna ecosystems, large ungulate herbivores are considered to be the major drivers of vegetation dynamics through directly reducing the abundance of the plants they consume and altering the competitive interactions between trees and grasses [28–30]. Intensive grazing by cattle is normally associated with an increase in woody vegetation [31], with wild browsing ungulates, such as elephants having the opposite effect [32]. Woody plants evolved with herbivory and herbivores play a key role in regulating their cover [33, 34]. In African savannas, herbivores include both invertebrates and vertebrates. Vertebrate herbivores range in size from the diks-diks (3–4 kg) to the elephant (6000 kg). The small herbivores tend to be selective concentrate feeders, whereas the large ones are bulk feeders because they cannot meet their daily feed requirements by being very selective [35]. Termites are an important group of herbivores as they can consume between 10 and 80 percent of available forage. The effects of herbivores on savanna ecosystems will vary depending on the vegetation type, the herbivore and the environment. Bond [25] found herbivory together with fire to be key determinants of vegetation structure and other ecosystem functions. Herbivores modify vegetation structure in many savanna ecosystems [36]. For example, browsers prevent woody plants recruitment to higher height strata [17, 32, 37]. This browser limitation of woody plant growth has been attributed either directly to browsing-induced mortality of woody seedlings and saplings or indirectly to fire, when browsing serves to suppress growth and maintain woody vegetation within the flame zone making them more susceptible to fire-induced mortality [29]. On the contrary, increases in woody cover have been attributed to overgrazing [31], which has been found to enhance dispersal of woody seeds, reduce competition from grazed grasses, reduce fire frequency and intensity due to lowered grass-fuel loads and increase water availability for deep-rooted woody plants as a result of lowered uptake by grasses [29, 31]. Sankaran et al. [17] reported higher woody cover in sites without elephants compared to those with high elephant biomass. Herbivory has both negative and positive effects on woody plants. For instance, megaherbivores (especially elephants) negatively affect woody plants [38], while intense herbivory by mesoherbivores increases woody plants density [39]. O'Connor [40] found elephants to kill woody plants mainly through complete uprooting. Additionally, herbivores enhance woody plant seed dispersal and increase germination rates following gut passage of the seeds, increasing the recruitment success of encroaching species [41, 42].

The major impact of herbivory, particularly by elephants, is to alter the structure and composition of vegetation by converting woodlands to shrublands and then to grasslands [32, 43]. Buechner and Dawkins [44] reported the conversion of *Terminalia glaucescens* woodlands, *Cynometra alexandri* rainforests and riparian woodlands to treeless grassland through the combined effects of elephants and fire in the Murchison Falls National Park, Uganda. Similar results have been reported from Tsavo National Park, Kenya, where elephants were shown to be the major cause of woodland decline and fire maintained the converted vegetation in a grassland state. Timberlake [45] reported that continuous browsing by elephant results in many small- and medium-sized trees being knocked down, effectively forming a shrubland 1.5–2 m high. Elephants break large trees resulting in an increase in shrub density from coppiced growth [46], with continued herbivory on shrubs preventing their recruitment into taller height classes [47]. Additionally, they can fell, push over or uproot trees and trample on seedlings [48, 49]. Elephants can fell as much as 20 percent of trees in an area per year, with the impacts more severe in restricted areas [50]. Woodland damage by elephants has been reported in the Kruger National Park in South Africa [51], the Luagwa Valley in Zambia [48], the Sengwa Wildlife Research area in northern Zimbabwe [52–54] and the savanna woodlands of East Africa [44, 55]. Furthermore, breakage of the main stems of trees results in a multi-stemmed growth form with limited vertical growth, altering woody vegetation structure [56].The multi-stemmed coppiced tree stems have high survivorship making them resilient to repeated herbivory which over time could lead to the development of a stable vegetation phase with low canopy cover but resistant to conversion into grassland [19, 56]. Eland also prevent recruitment of woody plants to higher height classes while at the same time causing extensive damage at lower height strata [57], while giraffe browsing reduces tree growth rates [58]. High impala densities have also been found to prevent the regeneration of Acacia tree populations through intense seedling predation [59]. Herbivory may lead to an increase in fast-growing palatable woody species or in slow-growing, often chemically defended, unpalatable species [60].

Fire and herbivory act synergistically in influencing woody plant density and composition [61]. Repeated herbivory exposes woody plants to fire by preventing their escape from the fire-prone lower height strata [62]. Additionally, elephants break or ring-bark large mature trees opening up their canopy, leading to an increase in herbage production in the woodlands, which in turn, increases the risk of intense annual fires that kill regenerating plants, converting woodland to shrubland or grassland. The interactive effects of elephants and fire have led to a decline of some woodlands and their subsequent replacement by grasslands or open savanna ecosystems [32, 63]. On the contrary, grazing herbivores through consumption of grass, reduce the fuel load, frequency and intensity of fires allowing woody plants to successfully establish [64]. Herbivores can also positively influence woody plant germination and establishment through other direct and indirect impacts such as trampling and seed dispersal [65]. Gordijn et al. [8] reported browsing as reducing the density of microphyllous palatable species which in turn were replaced by unpalatable macrophyllous species. Giraffe browsing has been found to result in extirpation of some deciduous microphyllous palatable species [66].

5. Effect of fire on woody plant communities in African savannas

Fire plays an important role in altering woody plant community structure in African savannas [4, 7]. It occurs in all savannas with most of the fires deliberately set by human beings, although there are some incidences of fire caused by lightning. Frequent fires reduce woody cover and maintain woody vegetation in a juvenile state by 'top-killing' seedlings and saplings, retarding transition to adulthood in tree species which can resprout from rootstocks after damage of aboveground structures [5, 22, 67]. In areas where fires have been suppressed, an increase in woody vegetation cover and density has been reported [8]. Woody cover is determined by tree abundance and size, with fire altering the population and community structure and tree size. Fire reduces the proportion of young trees that reach maturity, leading to a disproportionately large number of small trees [7]. In addition, fire reduces competition among mature trees leading to higher growth and survival rates. Repeated burning may result in bimodal tree size distribution, with small and large tree size classes being predominant. Fire also initiates processes such as coppicing which result in the production of multiple stems [7, 68]. Coppice regrowth is a strong regeneration response of woody species in the savanna [69].

Fire can destroy 50% or more of the annual forage production. Moist savannas produce high plant biomass, which in turn increase the fuel load resulting in intense fires. Conversely, herbivory causes a significant reduction in plant biomass accumulation thereby reducing the fuel load limiting the impacts of fires. However, elephant damage of trees makes them more susceptible to fire. Most tree damage occurs when fires are hot such as during late winter or early wet season as compared to mid-summer or wet seasons, with the hot early season burns damaging new plant growth. The frequency of burning also impacts on the extent of plant damage, with very frequent burns resulting in a reduction of plant biomass build up, thereby reducing the intensities of fire and the resulting damage to trees.

6. The role of resprouting in woody plant responses to disturbance

The abundance of woody vegetation in African savanna ecosystems is determined by their ability to respond to disturbance events. Disturbance events widely recognised to influence abundance of woody vegetation include fire [70], herbivory [71] and frost [72]. The ability of woody plants to resprout in response to disturbance events is important in sustaining woody plant populations, particularly in cases where seed production, germination and seedling survival are low [73]. Most woody plants in the savanna have the ability to resprout (coppice) and invest root reserves in rapid growth following a disturbance event [74, 75]. The removal of terminal shoots results in the breaking of apical dominance, allowing lateral meristems to develop into new shoots (hereafter referred to as resprouts) [76, 77]. The development of lateral buds into resprouts enables woody plants to tolerate persistent herbivory, through rapid replacement of lost photosynthetic tissue [78]. Resprouting is considered a strategy for the plant to produce cheap photosynthetic tissue to compensate for lost biomass and to quickly regain a positive carbon and nitrogen balance [77] and can be initiated from a root or stem [79].

The resprouting responses of woody plants to herbivory vary considerably. Choeni and Sebata [77] compared the resprouting abilities of five *Acacia* species in a semi-arid savanna by determining the number of resprouts following simulated herbivory. *Acacia karroo* was found to be a prolific resprouter, whereas *A. arenaria* produced very few resprouts (**Figure 1**).

The growth of resprouts following a disturbance event is very rapid to quickly replace lost photosynthetic plant material. For instance, sixfold resprout length increments within 10 weeks were reported in a study in central Zimbabwe [80]. Interestingly, three different woody species *viz. Grewia monticola* Sond., *Terminalia sericea* Burch. ex DC. and *Dichrostachys cinerea* (L.) Wight & Arn. have similar resprouting responses to disturbance (**Figure 2**).

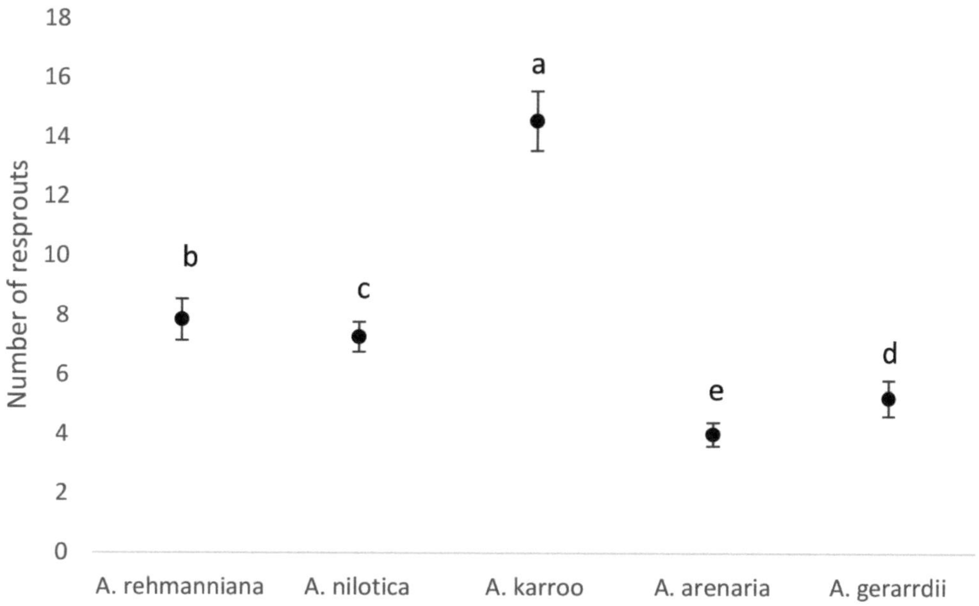

Figure 1. Mean (±SE) number of resprouts of *Acacia rehmanniana, A. nilotica, A. karroo, A. arenaria* and *A. gerrardii* in response to simulated herbivory in a semi-arid savanna. Source: Choeni and Sebata [77].

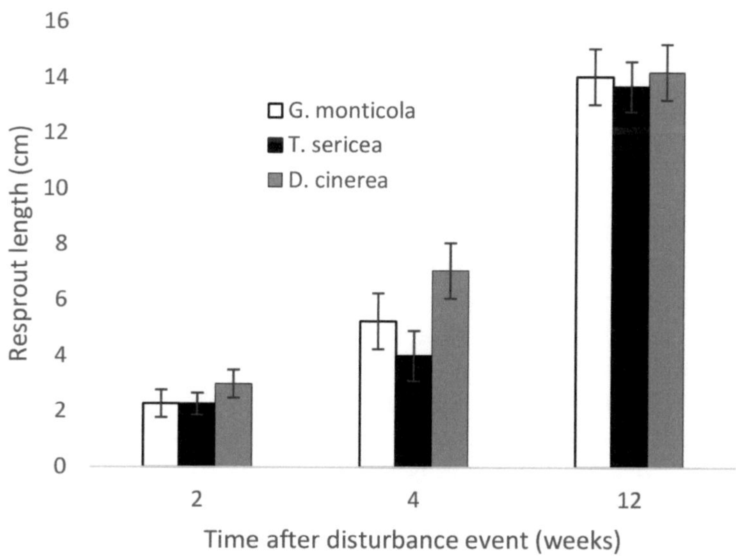

Figure 2. Mean (±SE) (n = 5) resprout lengths of three woody species following a disturbance event in a savanna ecosystem. Source: Huruba et al. [80].

Fornara and du Toit [78] found shoot growth rates to increase consistently with severity of herbivory. The timing of the disturbance event initiating resprouting will determine the resprout growth rate. For example, resprouting will be rapid when the plant growth conditions are favourable [81]. Sebata et al. [82] found resprout growth rates to be higher during the wet (growth) than the dry season in a southern African savanna. Resprouts benefit from better mobilization of stored energy reserves and higher photosynthetic rates during the growth season. Page and Whitham [83] found resprout growth to depend on the amount of carbohydrates that can be mobilized, by photosynthesis or in carbohydrate reserves.

7. Woody plant encroachment in African savannas

In most African ecosystems, open savannas are considered stable and productive because they are less prone to the rapid proliferation of new woody plants. This is due to the positive effect that large trees have on the natural functioning of the ecosystem which suppresses growth of woody plant seedlings and saplings. The open savanna is maintained through a process of system dynamics, which is based on the principle that the distance between a tree and its nearest neighbour of the same species is not determined purely by chance, but that tree spacing is normally distributed [4]. System dynamics predicts that the larger the tree, the greater is the distance between it and the nearest individual of the same species; this is particularly true for Acacia species. Reduced tree competition, through mortality, will result in increasing the growth rate of remaining individuals, whereas competition between individuals in a community will result in reduced growth in a tree population [4]. In the event that the system dynamics is upset, such as through loss of the established mature trees, there will be a rapid proliferation of woody plants, leading to an encroachment of the open savanna ecosystem. Woody plant encroachment is a common consequence of disturbance in savannas [84] and is characterised by an increase in density, cover, extent and biomass of trees in grass-dominated ecosystems [85]. It is a growing concern in most African savannas [86], negatively affecting cattle grazing, fuelwood provision, biodiversity conservation and ecosystem resilience [2]. Overgrazing, unsuitable fire regimes, increased carbon dioxide and climate change have been implicated as the key drivers of woody plant encroachment [29, 87, 88]. In African savannas, woody plant encroachment has generally been attributed to trees escaping from competition with grasses and browse pressure where cattle have replaced wildlife as the predominant herbivores [30]. Due to encroachment, an ecosystem transition takes place leading to an increase in shrub and tree cover in grasslands and savannas resulting in states of co-dominance by shrubs and grasses or complete conversion of grasslands to shrublands or tree-dominated woodlands [89]. These ecosystem transitions affect community composition and vegetation structure, ecosystem functions and biodiversity conservation [90]. However, some grasslands generate self-reinforcing mechanisms that promote conditions which prevent invasion by woody plants [89]. For instance, leaving little open space for colonization, producing many fibrous roots in the upper soil layers that can rapidly use water and nutrients and generating large amounts of herbaceous biomass that facilitates frequent and intense fires that kill unprotected woody plant meristems [91–93].

The competitive dynamics between grasses and woody plants changes once the later establish [89]. Woody plant seedlings recruit as single stems susceptible to top kill by fire then rapidly develop into multi-stemmed plants resistant to burning. Woody cover of the multi-stemmed plants increases limiting grass growth by reducing access to light. Eventually, grass cover, grass biomass and the fuel load decreases [94], allowing further woody plant recruitments [95]. Once woody plants establish themselves in grassland, it is difficult to reverse the process.

The most widely accepted theory explaining bush encroachment is the two-layer soil-water hypothesis [87, 96]. In this theory, the assumption is that water is the limiting factor and grasses use only topsoil moisture and nutrients, whereas woody plants use subsoil resources [97]. This separation of rooting niches allows for the coexistence of woody and grass plants [22]. A disruption of this relationship in favour of trees leads to woody plant proliferation. Overgrazing results in the grass roots extracting less water from the top soil layer and allowing more water to percolate into the sub-soil, where it is available for woody plant growth [22, 29]. Thus, overgrazing changes the grass-tree competitive interactions in favour of the woody plants and also reduces the fuel load through grass removal preventing woody plant damage by fire [67]. Grass competition restricts tree recruitment [39], although grasses may also have positive effects on tree establishment by protecting the saplings from mammalian browsers [98]. In open savannas, grasses generally outcompete trees for water and nutrients by growing fast and intercepting moisture from the upper soil layers, thereby preventing trees from gaining access to precipitation in the lower soil layers where their roots are mostly found [87]. Thus, when heavy grazing occurs, grasses are removed and soil moisture then becomes available to the trees, because they are more deeply rooted, allowing them to grow, recruit and expand [22]. Overgrazing of grasses has been identified as the main cause of increased woody plant density in the eastern areas of Botswana. Tree species with shallow roots (e.g. *Acacia mellifera* and *Grewia flava*) have been reported to be responsible for bush encroachment, suggesting that they are favoured by an increase in water availability in the surface soil following overgrazing of the grass layer. Heavy grazing also reduces the fuel load, which makes fires less intense and thus less damaging to trees and, consequently, results in an increase in woody vegetation. Woody plant encroachment in savannas can also be considered as a cyclical succession between open savanna and woody dominance that is driven by rainfall, which is highly variable, and inter-tree competition [22]. This means that savanna landscapes are composed of many patches in different states of transition between grass and woody dominance, that is, savannas are patch-dynamic systems. Alternatively, woody plant encroachment can be viewed as a natural recruitment process for savannas [22]. In recent years, the increasing carbon dioxide levels associated with global warming have been proposed to be favouring woody plant encroachment [73, 99]. Increases in atmospheric carbon dioxide improve water-use efficiency and increased carbon uptake in C_3 (mostly woody) plants favours them over C_4 (mostly grasses) [4, 73, 99]. The elevated carbon dioxide hypothesis is based on observations that most woody plants have the C_3 photosynthetic pathway and many of the grasses have the C_4 photosynthetic pathway. The C_3 photosynthetic pathway is advantageous at higher levels of carbon dioxide. Woody plant encroached areas can be converted back to open savannas through a process of self-thinning [22]. However, the interactive effects of tree growth with fire and herbivory make the process of self-thinning complex and prolonged.

8. Effect of land degradation on woody plant communities and opportunities for restoration

Rangeland degradation starts with the formation of small bare patches which then expand to form large denuded areas [100], leading to the reduction or loss of biodiversity and woody plant productivity [101]. The recovery of woody vegetation in severely degraded areas through natural succession processes is very slow, necessitating active intervention in the form of restoration efforts [102]. Restoration efforts are aimed at returning ecosystems to their previously stable states through the re-establishment of lost vegetation. Cairns [103] argued that restoration must be firmly rooted in ecology and the connection between ecological succession and ecological restoration. The presence of large trees in an African savanna represents a structured stable ecosystem that is productive because of the benefits of the presence of woody plants such as soil enrichment, favourable sub-habitats for the maintenance of positive grass-tree associations and increased stability as large trees may suppress the establishment and development of woody seedlings under their canopies or in their close proximity [4]. The presence of large trees plays a critical role in the restoration of degraded rangelands, particularly the suppression of bush encroachment. Smit [4] suggested that restoring savanna structure requires a highly selective approach where woody plants are thinned in such a way that the remaining trees will benefit from the reduced competition from other woody plants, resulting in increased growth and thus an increasing sphere of influence on newly establishing seedlings.

In most African savannas, woody plant recovery is facilitated through the initial development of pioneer woody species, usually xerophytic spinescent microphyllous species, which are then replaced by a more stable savanna consisting of long-lived broad-leaved species [4]. The broad-leaved woody species are able to germinate and develop under the canopies of spinescent species like some *Acacia* species. The spinescent species later succumb to natural causes, enabling the broad-leaved species to predominate, as the former are unable to establish under the canopies of established trees [4].

The vegetation structure of African savannas is being altered by expansion of human settlements into previously undisturbed areas [104], with fuelwood harvesting as the key driver [105]. Preventing loss of plant communities is more cost-effective than attempts to restore degraded ecosystems, because restoration processes will require costly inputs, such as woody plant establishment. Nonetheless, restoration of degraded savanna systems is inevitable to increase rangeland productivity. Restoration measures include the re-introduction of desired grass species or other investments to improve rangeland quality from both an ecological and an economic perspective.

Attempts at restoring encroached areas by the removal of some or all of the woody plants will normally result in an increase of grass production and thus also the grazing capacity. However, the rapid establishment of tree seedlings after the removal of some or all of the mature woody plants may reduce the effective time span of restoration measures. In many cases, the resultant re-establishment of new woody seedling may in time develop into a state that is worse than the original state. To counter this, a more stable environment can be created by maintaining or restoring savanna structure (large trees). In a structured savanna, large

http://dx.doi.org/10.5772/intechopen.69865

trees are able to suppress the establishment of new seedlings, while maintaining the other benefits of woody plants like soil enrichment and the provision of food to browsing herbivore species. The loss of large trees from savanna ecosystems through indiscriminate, non-selective bush control measures results in failure to successfully restore encroached areas. Restoration should involve a highly selective process where woody plants are thinned in such a way that the remaining trees will benefit from the reduced competition from other woody plants. Direct competition between grasses and woody seedlings and saplings for soil water can suppress the recruitment of woody species.

9. Future perspective

Woody plants in African savanna ecosystems are increasingly being cleared to make way for cropping or being harvested as poles for construction of houses as human population increases. The loss of woody plants alters the ecology of African savanna ecosystems affecting the goods and services offered. A better understanding of the ecology of woody plants is essential in managing these ecosystems in addition to attempts to restore severely degraded areas. Further research to better understand the interactions between the woody and herbaceous components of the African savannas is required. The phenomenon of bush encroachment remains poorly understood necessitating further research to gain better insights.

10. Conclusion

A better understanding of the ecology of woody plants is a key to the sustainable management of African savannas because they stabilize these ecosystems. Rainfall, soil, herbivory and fire play important roles in shaping African savanna ecosystems. The resilience and persistence of woody plants in African savannas are determined by their ability to resprout quickly after a disturbance event. Woody plant encroachment disrupts the balance between woody and herbaceous plants resulting in the replacement of the more productive open savannas with less productive densely vegetated ecosystems. Land degradation, mainly due to overgrazing, accelerates woody plant encroachment. Successful restoration of degraded savanna ecosystems requires a good understanding of their ecology.

Author details

Allan Sebata

Address all correspondence to: allan.sebata@nust.ac.zw

Department of Forest Resources and Wildlife Management, Faculty of Applied Sciences, National University of Science and Technology, Zimbabwe

References

[1] Frost PG, Medina E, Menaut JC, Solbrig O, Swift M, Walker BH. Response of Savannas to Stress and Disturbance. Biology International Special Issue 10, Paris: IUBS; 1986

[2] Fisher JT, Witkowski ETF, Erasmus BFN, Mograbi PJ, Asner GP, van Aardt JAN, Wessels KJ, Mathieu R. What lies beneath: Detecting sub-canopy changes in savanna woodlands using a three-dimensional classification method. Applied Vegetation Science. 2015;**18**:528-540

[3] House JI, Archer S, Breshears DD, Scholes RJ. Conundrums in mixed woody–herbaceous plant systems. Journal of Biogeography. 2003;**30**:1763-1777

[4] Smit GN. An approach to tree thinning to structure southern African savannas for long-term restoration from bush encroachment. Journal of Environmental Management. 2004;**71**:179-191

[5] Bond WJ, Woodward FI, Midgley GF. The global distribution of ecosystems in a world without fire. New Phytologist. 2005;**165**:525-537

[6] Sankaran M, Hanan NP, Scholes RJ, Ratnam J, Augustine DJ, Cade BS, Gignoux J, Higgins SI, Le Roux X, Ludwig F, Ardo J, Banyikwa F, Bronn A, Bucini G, Caylor KK, Coughenour MB, Diouf A, Ekaya W, Feral CJ, February EC, Frost PGH, Hiernaux P, Hrabar H, Metzger KL, Prins HHT, Ringrose S, Sea W, Tews J, Worden J, Zambatis N. Determinants of woody cover in African Savannas. Nature. 2005;**438**:846-849

[7] Devine AP, Stott I, McDonald RA, Maclean IMD. Woody cover in wet and dry African savannas after six decades of experimental fires. Journal of Ecology. 2015;**103**:473-478

[8] Gordijn PJ, Rice E, Ward D. The effects of fire on woody plant encroachment are exacerbated by succession of trees of decreased palatability. Perspectives in Plant Ecology, Evolution and Systematics. 2012;**14**:411-422

[9] Scholes RJ, Bond WJ, Eckhardt HC. Vegetation dynamics in the Kruger ecosystem. In: du Toit JT, Rogers KH, Biggs HC, editors. The Kruger Experience. Washington: Island Press; 2003. pp. 242-262

[10] Scholes R, Dowty PR, Caylor K, Parsons DAB, Frost PGH, Shugart HH. Trends in savanna structure and composition along an aridity gradient in the Kalahari. Journal of Vegetation Science. 2002;**13**:419-428

[11] Mucina L, Rutherford MC. The Vegetation of South Africa, Lesotho and Swaziland. Strelitzia 19. Pretoria: South African National Biodiversity Institute; 2006

[12] O'Brien EM. Climatic gradients in woody plant species richness: Towards an explanation based on an analysis of southern Africa's woody flora. Journal of Biogeography. 1993;**20**:181-198

[13] Huntley BJ, Walker BH, editors. Ecology of tropical savannas. Berlin, DE: Springer-Verlag; 1982

[14] Scholes RJ, Walker BH. An African Savanna: Synthesis of the Nylsvley Study. Cambridge: Cambridge University Press; 1993

[15] Mills AJ, Milewski AV, Fey MV, Gröngröft A, Petersen A, Sirami C. Constraint on woody cover in relation to nutrient content of soils in western southern Africa. Oikos. 2013;**122**:136-148

[16] Kraaij T, Ward D. Effects of rain, nitrogen, fire and grazing on tree recruitment and early survival in bush-encroached savanna, South Africa. Plant Ecology. 2006;**86**:235-246

[17] Sankaran M, Ratnam J, Hanan N. Woody cover in African savannas: The role of resources, fire and herbivory. Global Ecology and Biogeography. 2008;**17**:236-245

[18] Childes SL, Walker BH. Ecology and dynamics of the woody vegetation on the Kalahari sands in Hwange National Park, Zimbabwe. Vegetation. 1987;**72**:111-128

[19] Holdo RM. Elephant herbivory, frost damage and topkill in Kalahari sand woodland savanna trees. Journal of Vegetation Science. 2006;**17**:509-518

[20] Campbell BM, Cunliffe RN, Gambiza J. Vegetation structure and small-scale pattern in Miombo woodland, Marondera, Zimbabwe. Bothalia. 1995;**25**:121-126

[21] Cole MM. The influence of soils, geomorphology and geology on the distribution of plant communities in savanna ecosystems. In: Huntley BJ, Walker BH. editors, Ecology of Tropical Savannas. Ecological Studies. Berlin: Springer; 1982.Vol. **24**. pp. 145-174

[22] Wiegand K, Saltz D, Ward D. A patch-dynamics approach to savanna dynamics and woody plant encroachment – Insights from an arid savanna. Perspectives in Plant Ecology, Evolution and Systematics. 2006;**7**:229-242

[23] Scholes RJ. Savanna. In: Cowling RM, Richardson DM, Pierce SM. editors, Vegetation of Southern Africa. Cambridge, MA: Cambridge University Press; 1997.pp. 258-277

[24] Walker BH, Langridge JL. Predicting savanna vegetation structure on the basis of plant available moisture and plant available nutrients: A case study from Australia. Journal of Biogeography. 1997;**24**:813-825

[25] Bond WJ. What limits trees in C_4 grasslands and savannas? Annual Review of Ecology, Evolution, and Systematics. 2008;**39**:641-659

[26] Williams RJ, Duff GA, Bowman DMJS, Cook GD. Variation in the composition and structure of tropical savannas as a function of rainfall and soil texture along a large scale climatic gradient in the Northern Territory, Australia. Journal of Biogeography. 1996;**23**:747-756

[27] Scholes RJ, 1990. The influence of soil fertility on the ecology of southern African dry savannas. Journal of Biogeography. 1990;**17**:415-419

[28] Skarpe C. Dynamics of savanna ecosystems. Journal of Vegetation Science. 1992;**3**:293-300

[29] Scholes RJ, Archer SR. Tree-Grass Interactions in Savannas. Annual Review of Ecology and Systematics. 1997;**28**:517-544

[30] Riginos C, Young TP. Positive and negative effects of grass, cattle, and wild herbivores on Acacia saplings in an East African savanna. Oecologia. 2007;**153**:985-995

[31] Roques KG, O'Connor TG, Watkinson AR. Dynamics of shrub encroachment in an African savanna: Relative influences of fire, herbivory, rainfall and density dependence. Journal of Applied Ecology. 2001;**38**:268-280

[32] Dublin HT, Sinclair ARE, Mcglade J. Elephants and fire as causes of multiple stable states in the Serengeti-Mara woodlands. Journal of Animal Ecology. 1990;**59**:1147-1164

[33] Cumming DHM. The influence of large herbivores on savanna structure in Africa. In: Huntley BJ and Walker BH, editors. Ecology of Tropical Savannas. Springer-Verlag, Berlin, Germany; 1982. pp. 217-245

[34] Holdo RM, Holt RD, Fryxell JM. Herbivore–vegetation feedbacks can expand the range of savanna persistence: Insights from a simple theoretical model. Oikos. 2013;**122**:441-453

[35] Hofmann RR. Evolutionary steps of ecophysiological adaptation and diversification of ruminants: A comparative view of their digestive system. Oecologia. 1989;**78**:443-457

[36] Staver AC, Bond WJ, Stock WD, Van Rensburg SJ, Waldram MS. Browsing and fire interact to suppress tree density in an African savanna. Ecological Applications. 2009;**19**:1909-1919

[37] Augustine DJ, McNaughton SJ. Regulation of shrub dynamics by native browsing ungulates on East African rangeland. Journal of Applied Ecology. 2004;**41**:45-58

[38] Owen-Smith N. Pleistocene extinctions: The pivotal role of megaherbivores. Paleobiology. 1987;**13**:351-362

[39] Goheen JR, Palmer TM, Keesing F, Riginos C, Young TP. Large herbivores facilitate savanna tree establishment via diverse and indirect pathways. Journal of Animal Ecology. 2010;**79**:372-382

[40] O'Connor TG. Demography of woody species in a semi-arid African savanna reserve following the re-introduction of elephants. Acta Oecologica. 2017;**78**:61-70

[41] Miller MF. Acacia seed survival, seed-germination and seedling growth following pod consumption by large herbivores and seed chewing by rodents. African Journal of Ecology. 1995;**33**:194-210

[42] Tews J, Schurr F, Jeltsch F. Seed dispersal by cattle may cause shrub encroachment of *Grewia lava* on southern Kalahari rangelands. Applied Vegetation Science. 2004;**7**:89-102

[43] Ben-Shahar R. Changes in structure of savanna woodlands in northern Botswana following the impacts of elephants and fire. Plant Ecology. 1998;**136**:189-194

[44] Buechner HK, Dawkins HC. Vegetation change induced by elephants and fire in Murchison Falls National Park, Uganda. Ecology. 1961;**42**:752-766

[45] Timberlake JR. *Colophospermum mopane,* Annotated Bibliography and Review. Harare: The Zimbabwe Bulletin of Forestry Research No. 11, Forestry Commission; 1995

[46] Richardson-Kageler SJ. Large mammalian herbivores and woody plant species diversity in Zimbabwe. Biodiversity and Conservation. 2003;**12**:703-715

[47] Gadd ME. The impact of elephants on the marula tree (*Sclerocarya birrea*). African Journal of Ecology. 2002;**40**:328-336

[48] Caughley G. The elephant problem—An alternative hypothesis. East African Wildlife Journal. 1976;**14**:265-283

[49] Mapaure I, Mhlanga L. Elephants and woodlands: The impact of elephant damage on *Colophospermum mopane* on Namembere Island, Lake Kariba, Zimbabwe. Zimbabwe Science News. 1998;**32**:15-19

[50] Conybeare AM. Elephant occupancy and vegetation change in relation to artificial water points in Kalahari sand area of Hwange National Park. Unpublished MSc thesis, Department of Biological Sciences, University of Zimbabwe. Harare, Zimbabwe; 1991

[51] van Wyk P, Fairall N. The influence of the African elephant on the vegetation of Kruger National Park. Koedoe. 1969;**9**:57-95

[52] Anderson GD, Walker BH. Vegetation composition and elephant damage in the Sengwa Research Area, Rhodesia. Journal of South African Wildlife Research Management Association. 1974;**4**:1-14

[53] Guy PR. The feeding behaviour of elephant (*Loxodonta africana*) in the Sengwe area, Rhodesia. South African Journal of Wildlife. 1976;**6**:55-63

[54] Guy PR. The influence of elephants and fire on a *Brachystegia-Julbernardia* woodland in Zimbabwe. Journal of Tropical Ecology. 1989;**5**:215-226

[55] Pellew RAP. The impacts of elephant, giraffe and fire upon the *Acacia–Tortilis* woodlands of the Serengeti. African Journal of Ecology. 1983;**21**:41-74

[56] Holdo RM. Woody plant damage by African elephants in relation to leaf nutrients in western Zimbabwe. Journal of Tropical Ecology. 2003;**19**:189-196

[57] Nyengera R, Sebata A. Effect of eland density and foraging on *Combretum apiculatum* physiognomy in a semi-arid savanna. African Journal of Ecology. 2009;**48**:45-50

[58] Birkett A, Stevens-Wood B. Effects of low rainfall and browsing by large herbivores on an enclosed savanna habitat in Kenya. African Journal of Ecology. 2005;**43**:123-130

[59] Prins HHT, van der Jeugd HP. Herbivore population crashes and woodland structure in East Africa. Journal of Ecology. 1993;**81**:305-314

[60] Skarpe C, Aarrestad PA, Andreassen HP, Dhillion SS, Dimakatso T, du Toit JT, Halley DJ, Hytteborn H, Makhabu S, Mari M, Marokane W, Masunga G, Modise D, Moe SR, Mojaphoko R, Mosugelo D, Motsumi S, Neo-Mahupeleng G, Ramotadima M, Rutina, L, Sechele L, Sejoe TB, Stokke S, Swenson JE, Taolo C, Vandewalle M, Wegge P. The return of the giants: Ecological effects of an increasing elephant population. Ambio. 2004;**33**:276-282

[61] Midgley JJ, Lawes MJ, Chamaillé-Jammes S. Savanna woody plant dynamics: The role of fire and herbivory, separately and synergistically. Australian Journal of Botany. 2010;**58**:1-11

[62] Jachmann H, Bell RHV. Utilization by elephants of the Brachystegia woodlands of the Kasungu National Park, Malawi. African Journal of Ecology. 1985;**23**:245-258

[63] Laws RM. Elephants as agents of habitat and landscape change in East Africa. Oikos. 1970;**21**:1-15

[64] van Langevelde F, van de Vijver CADM, Kumar L, van de Koppel J, de Ridder N, van Andel J, Skidmore AK, Hearne JW, Stroosnijder L, Bond WJ, Prins HHT, Rietkerk M. Effects of fire and herbivory on the stability of savanna ecosystems. Ecology. 2003;**84**:337-350

[65] Danell K, Duncan P, Bergström R, Pastor J. Large Herbivore Ecology, Ecosystem Dynamics and Conservation. Cambridge, UK: Cambridge University Press; 2006

[66] Bond WJ, Loffell D. Introduction of giraffe changes acacia distribution in a South African savanna. African Journal of Ecology. 2001;**39**:286-294

[67] Midgley JJ, Bond WJ. A synthesis of the demography of African acacias. Journal of Tropical Ecology. 2001;**17**:871-886

[68] Chidumayo EN. Growth responses of an African savanna tree, *Bauhinia thonningii* Schumacher, to defoliation, fire and climate. Trees. 2007;**21**:231-238

[69] Kaschula S, Twine W, Scholes M. The effect of catena position and stump characteristics on the coppice response of three savannah fuelwood species. Environmental Conservation. 2005;**32**:76-84

[70] Bond WJ. Fire. In: Cowling RM, Richardson DM, Pierce SM, editors. Vegetation of Southern Africa. Cambridge: Cambridge University Press; 1997

[71] O'Connor TG, Puttick JR, Hoffman MT. Bush encroachment in southern Africa: Changes and causes. African Journal of Range and Forage Science. 2014;**31**:67-88

[72] Muller K, O'Connor TG, Henschel JR. Impact of a severe frost event in 2014 on woody vegetation within the Nama-Karoo and semi-arid savanna biomes of South Africa. Journal of Arid Environments. 2016;**133**:112-121

[73] Bond WJ, Midgley GF. A proposed CO_2-controlled mechanism of woody plant invasion in grasslands and savannas. Global Change Biology. 2000;**6**:865-869

[74] Bergström R. Browse characteristics and impact of browsing on trees and shrubs in African savannas. Journal of Vegetation Science. 1992;**3**:315-324

[75] Whitecross MA, Archibald S, Witkowski ETF. Do freeze events create a demographic bottleneck for *Colophospermum mopane*? South African Journal of Botany.2012;**83**:9-18

[76] Aarssen LW. Hypothesis for the evolution of apical dominance in plants: Implications for the interpretation of overcompensation. Oikos. 1995;**74**:149-156

[77] Choeni H, Sebata A. Interspecific variation in the resprouting responses of *Acacia* species following simulated herbivory in a semi-arid southern African savannah. African Journal of Ecology. 2014;**52**:479-483

[78] Fornara DA, du Toit JT. Browsing lawns? Responses of *Acacia nigrescens* to ungulate browsing in an African savanna. Ecology. 2007;**88**:200-209

[79] Bond WJ, Midgley JJ. The evolutionary ecology of sprouting in woody plants. International Journal of Plant Sciences. 2003;**164**:S103-S114

[80] Huruba R, Mundy PJ, Sebata A, Purchase GK, MacFadyen DN. Impala, *Aepyceros melampus*: does browse quality influence their use of sites originally utilised as short-duration kraals in a southern African savanna? The Rangeland Journal. 2017;**39**:113-121; in press. http://dx.doi.org/10.1071/RJ16016

[81] Strauss SY, Agrawal AA. The ecology and evolution of plant tolerance to herbivory. Trends in Ecological Evolution. 1999;**14**:179-185

[82] Sebata A, Nyathi P, Mlambo D. Growth responses of *Grewia flavescens* Juss. (Sandpaper Raisin) and *Grewia monticola* Sond. (Grey Grewia) (Tiliaceae) to shoot clipping in a semi-arid Southern African savanna. African Journal of Ecology. 2009;**47**:794-796

[83] Page KN, Whitham TG. Overcompensation in response to mammalian herbivory: The advantage of being eaten. American Naturalist. 1987;**129**:407-416

[84] Vadigi S, Ward D. Herbivory effects on saplings are influenced by nutrients and grass competition in a humid South African savanna. Perspectives in Plant Ecology, Evolution and Systematics. 2014;**16**:11-20

[85] Archer S, Scifres C, Bassham CR, Maggio R. Succession in a subtropical savanna: Conversion of grassland to thorn woodland. Ecological Society of America. 1988;**58**: 111-127

[86] Archer S, Schimel DS, Holland EA. Mechanisms of shrubland expansion: Land use, climate or CO_2? Climatic Change. 1995;**29**:91-99

[87] Walker BH, Ludwing D, Holling CS, Peterman RM. Stability of semi-arid savanna grazing systems. Journal of Ecology. 1981;**69**:473-498

[88] Bond WJ, Midgley GF. Carbon dioxide and the uneasy interactions of trees and savannah grasses. Philosophical Transactions of the Royal Society. 2012;**367**:601-612

[89] Ratajczak Z, Nippert JB, Briggs JM, Blair JM. Fire dynamics distinguish grasslands, shrublands and woodlands as alternative attractors in the Central Great Plains of North America. Journal of Ecology. 2014;**102**:1374-1385

[90] Ratajczak Z, Nippert JB, Collins SL. Woody encroachment decreases diversity across North American grasslands and savannas. Ecology. 2012;**93**:697-703

[91] Briggs JM, Knapp AK. Determinants of C_3 forb growth and production in a C_4 dominated grassland. Plant Ecology. 2001;**152**:93-100

[92] Nippert JB, Knapp AK. Linking water uptake with rooting patterns in grassland species. Oecologia. 2007;**153**:261-272

[93] van Wilgen BW, Trollope WSW, Biggs HC, Potgieter ALF, Brockett BH. Fire as a driver of ecosystems variability. The Kruger Experience. In: Du Toit JT, Rogers KH, Biggs HC, editors. Ecology and Management of Savanna Heterogeneity. Washington, DC: Island Press; 2002. pp. 149-170

[94] Heisler JL, Briggs JM, Knapp AK, Blair JM, Seery A. Direct and indirect effects of fire on shrub density and above-ground diversity in a mesic grassland. Ecology. 2004;**85**: 2245-2257

[95] Ratajczak Z, Nippert JB, Hartman JC, Ocheltree TW. Positive feedbacks amplify rates of woody encroachment in mesic tallgrass prairie. Ecosphere. 2011;**2**:1-14

[96] Walter H. Ecology of Tropical and Subtropical Vegetation. Edinburgh: Oliver and Boyd; 1971

[97] Knoop WT, Walker BH. Interactions of woody and herbaceous vegetation in a southern African savanna. Journal of Ecology. 1985;**73**:235-253

[98] Seymour C. Grass, rainfall and herbivores as determinants of *Acacia erioloba* (Meyer) recruitment in an African savanna. Plant Ecology. 2008;**197**:131-138

[99] Ward D. A resource ratio model of the effects of changes in CO_2 on woody plant invasion. Plant Ecology. 2010;**209**:147-152

[100] Kellner K, Bosch OJH. Influence of patch formation in determining the stocking rate for southern African grasslands. Journal of Arid Environments. 1992;**22**:99-105

[101] Okin G. Desertification: Monitoring and Forecasting, Desertification—Introduction; 2002. website: http://www.planetary.caltech.edu/_arid/desert/desert.html

[102] van den Berg L, Kellner K. Restoring degraded patches in a semi-arid rangeland of South Africa. Journal of Arid Environments. 2005;**61**:497-511

[103] Cairns J. Rehabilitating Damaged Ecosystems. 2nd ed. Boca Raton: Lewis Publishers; 1995

[104] Coetzer KL, Erasmus BFN, Witkowski ETF, Bachoo AK. Land cover change in the Kruger to Canyons Biosphere Reserve (1993-2006): A first step towards creating a conservation plan for the subregion. South African Journal of Science. 2010;**106**:26-35

[105] Matsika R, Erasmus BFN, Twine W. A tale of two villages: Assessing the dynamics of fuelwood supply in communal landscapes in South Africa. Environmental Conservation. 2013;**40**:71-83

Modification in Grassland Ecology under the Influence of Changing Climatic and Land Use Conditions

Jimin Cheng, Chengcheng Gang, Liang Guo, Wei Li,
Jingwei Jin, Jishuai Su and Lin Wei

Additional information is available at the end of the chapter

http://dx.doi.org/10.5772/intechopen.69478

Abstract

Grasslands are important terrestrial ecosystems in China, which are mainly distributed in arid and semiarid regions. Based on the multiyear field experiments in the semiarid grassland, the effects of land use practices on grassland above- and belowground community characteristics were investigated. In addition, how the annual climate factors regulate grassland productivity was also studied to detect critical periods for grass growth. Results showed that grazing exclusion increased grassland root biomass, root length density and root surface area with declining plant species richness. After grazing exclusion, with perennial bunchgrasses being predominant in root community all the time, proportion of perennial rhizome grasses increased and proportion of perennial forbs declined. Clipping significantly decreased the annual mean soil respiration and its components. The root respiration was more sensitive to clipping than microbial respiration. Temperature increments during the early stage of the growing season (April–May) were positively correlated with aboveground productivity. However, hot and dry summer (June–July) strongly inhibited aboveground productivity. Impacts of drought and heat in August on productivity were negligible. Increased temperature and precipitation during the senescence period (September–October) and a warmer dormancy phase (November–March) were negatively correlated with productivity in the following year, while precipitation during the dormancy period had no detectable effects.

Keywords: semiarid grassland, grazing exclusion, soil respiration, climate variation, biodiversity, productivity

1. Introduction

Grasslands are among the largest biomes in the world, accounting for nearly 25% of the land surface on earth [1, 2]. Grassland ecosystem plays a key role in balancing the concentrations of global atmospheric greenhouse gases through carbon storage and sequestration [3]. Grasslands also significantly contribute to food security by providing food for ruminants, which are sources of meat and milk for human consumption. China has nearly 4 million km^2 of grasslands, accounting for 40% of China's total land area and 13% of the world's total grassland [4, 5]. Concurrent with population growth and socioeconomic development, however, China's grasslands have experienced rapid degradation over the last few decades due to climate change and unsound anthropogenic impacts [6, 7]. To combat the grassland degradation and restoration of the environment, the Chinese government has launched batches of national-scale conservation policies during the late 1990s and early 2000s. Two of them, the Grain for Green Program (GGP) and the Grazing Withdrawal Program (GWP) cover most of the grassland regions [8–10]. Restoring degraded grassland ecosystems is critical to the ecological and economical sustainability of these systems.

About 90% of grassland was degraded as a consequence of overgrazing by livestock in China [11]. Overgrazing induced considerable destructive effects on plant community and soil resources [12]. Grazing exclusion has been proven to be a successful practice to restore degraded grasslands throughout the world [13, 14]. Many studies pointed out significant enhancing effects of grazing exclusion on plant coverage, density and aboveground biomass in the early stage, which were diluted or even reversed as grazing exclusion time increased [11, 15]. Meanwhile, grazing exclusion not only significantly increased storage and availability of soil water and nutrients through more litter inputs [14, 16], but also played an important role in structuring community of soil eukaryotes [17]. Contrasted with numerous researches on aboveground responses to grazing exclusion, researches about root responses are largely limited by the studying difficulties and complexity of plant roots. Current studies on fenced grassland root mainly focused on root biomass and its distribution pattern in different types of grassland [12, 18]. Root morphology and/or physiology traits and plasticity have received considerable attentions due to their capability of foraging soil nutrients [19, 20]. There is a considerable difference in root traits and plasticity among different plant species, normally with greater ones in graminaceous species [21]. The hierarchy of root trait values and plasticity among species and plant functional groups in the vegetation could drive early-stage competition for water and nutrients, which ultimately made an effect on vegetative succession [22, 23]. However, major knowledge gaps still exist, concerning responses of plant root morphological traits and root community composition to grazing exclusion in long-term restored grassland.

Soil respiration plays an important role in regulating soil C pools and net C balance in terrestrial ecosystems [24]. The rate of soil respiration can be influenced by climate change (global warming, precipitation regimes, etc.), as well as anthropogenic activity (land use change and management practice), with consequent impacts on terrestrial C cycling and feedbacks to climate change [25, 26]. As one of the common land use practice, clipping or mowing of hay is regarded as a critical component of global change [27]. The effect of clipping on soil respiration had been investigated widely in different ecosystems; however, the results were various and inconsistent with each other [28, 29]. One reason for the variability of previous studies in clipping effect on soil respiration is that soil respiration is composed of two different components. One of the components is root respiration, which refers to the CO_2 emission from plant roots, mycorrhizal fungi and other associated microorganisms (rhizosphere microorganisms) that depend on the contemporaneous [30]. Another component is microbial respiration, which is defined as the CO_2 emission from the decomposition of plant litter and soil organic matter by soil microorganisms [31]. Substrate sources of the two soil respiration components have different magnitudes, turnover rates and seasonal patterns, which make the two soil respiration components respond differently to climate change and land use practice [27, 32]. In addition, the contributions of root respiration to soil respiration are various in different ecosystems, which may also be responsible for the inconsistent results of the clipping effect on soil respiration [33]. Hence, quantifying the individual changes of root and microbial respiration in response to clipping is imperative for a comprehensive understanding of ecosystem carbon cycling.

Climate-driven variability in grassland productivity impacts the global carbon balance, ecosystem service delivery, and profitability of pastoral livelihoods. Aboveground net primary productivity (ANPP) of grasslands is highly temporally variable, as compared to other ecosystems, such as forest and cropland [34]. Much of the previous work considering the impacts of climate variability on ANPP has focused on annual precipitation and temperature [35, 36]. While the importance of these annual-scale metrics has often been confirmed in studies at regional scales, numerous site-specific reports have indicated that inter-annual variability in ANPP is poorly or even not at all correlated with annual climate conditions [37], with much of the temporal variation in ANPP left unexplained [36]. Changes in precipitation or temperature during certain parts of the year are more relevant drivers of ANPP than annual changes [38, 39], since vegetation production responds differently to climatic variation during different seasons [38, 40]. Future climates are likely to include more frequent extreme weather events and more pronounced seasonal variation in temperature and precipitation.

To provide a new perspective of biodiversity restoration and the basis for management of degraded grassland in semiarid areas, we firstly conducted with a space for time substitution method at Yunwushan National Nature Reserve, a typical steppe grassland on the Loess Plateau with different grazing exclusion timescales to determine effects of grazing exclusion on grassland root biomass, morphological traits and root community compositions in plant functional group level. Then, a clipping experiment was carried out to investigate the effect of clipping on root and microbial respiration. Finally, long-term productivity and weather records since 1982 were collected to examine the impacts of climate variability at different times of the year on grassland productivity.

2. Materials and methods

2.1. Study area

This research was conducted in Yunwushan National Natural Grassland Protection Zone in Ningxia Hui Autonomous Region, China (36°10′-36°17′N, 106°21′-106°27′E, 1800–2100 m a.s.l.). Since 1982, the grassland has had been protected as a long-term monitoring sites for restoration of degraded grassland. The site is located at an elevation from 1800 to 2100 m and has a total area of 6660 ha. Mean annual temperature during 1982–2011 was 7°C with mean monthly temperature extremes of −22°C in January and 25°C in July. Annual precipitation averaged 425 mm. Annual evaporation is 1017–1739 mm, and the frost-free season averages 137 days. Soil type in the study area is montane gray-cinnamon soil. The vegetation community consists of 297 plant species and is dominated by *Stipa* plants (*Stipa bungeana, Spectrunculus grandis, Salvia przewalskii*), and main forbs include *Artemisia sacrorum* and *Thymus mongolicus*.

2.2. Experimental design and sampling

2.2.1. *Grazing exclusion*

Five experimental sites along a chrono-sequence of grassland restoration were selected in August 2012, when peak aboveground biomass occurred, with grazing exclusion for 30 years (GE30), 22 years (GE22), 9 years (GE09), 5 years (GE05) and continuous grazing at a medium density during the whole year (four sheep/ha) (GG), respectively. A transect of 300 × 100 m with representative vegetation was selected as the study area within each site, in which three pseudo-replicated plots (30 × 30 m) were established, and three subplots (2 × 2 m) were set up with a minimum interval of 15 m in each plot for field sampling.

2.2.2. *Soil sampling*

With aboveground plant parts being attached, a soil block of 50 cm long × 50 cm wide × 30 cm deep was excavated in each subplot and then was gently loosen by hand to get the intact root-soil mixtures with minimal breakage. Plant root-soil mixtures were soaked in water for twenty minutes and were gently shaken for several times to remove bulk soil.

2.2.3. *Plant root sampling*

Plant roots were carefully washed under flowing water to remove tightly attached organic matter and mineral soils and carefully identified roots in plant functional group level according to plant aboveground parts, root color, diameter, branches and texture. Five functional groups (PFGs) were categorized as perennial rhizome grass (PR), perennial bunchgrass (PB), PF perennial forbs (PF), shrubs and semishrubs (SS) and annuals and biennials (AB) [26, 41]. Functional group richness and species richness were the number of functional groups and plant root species appearing in one subplot, respectively.

After cutting down plant aboveground parts, roots in the same plant functional groups were spread on a transparent, plastic tray and scanned at a resolution of 300 dpi (Epson Scanner (10000XLPro, Canada). Root images were analyzed with WinRhizoPro software (V2012b, Regent Instruments, Canada) to measure root length (m), root surface area (cm^2). Thereafter, roots were oven-dried at 65°C for 48 h and then weighed to gain root mass. Root biomass, root length density, specific root length and specific root area are calculated in equations [42] as follows:

$$Root\ biomass\ (RB,\ g\ m^{-2}) = Root\ mass/Sampling\ area \quad (1)$$

$$Root\ length\ density\ (RLD,\ mm^{-3}) = Root\ length/Sampling\ volume \quad (2)$$

$$Specific\ root\ length\ (SRL,\ mg^{-1}) = Root\ length/Root\ mass \quad (3)$$

$$Specific\ root\ area\ (SRA,\ cm^2\ g^{-1}) = Root\ surface\ area/Root\ mass \quad (4)$$

2.2.4. Clipping management

The experiment was designed as a randomized block with five replicate blocks. Clipping was done once a year in the spring (June 20, 2014, and June 16, 2015). The trenching method was used in this study to separate soil respiration into root and microbial respiration [43]. In each plot, one root-free small plot (0.3 × 0.3 m) lined with nylon mesh (0.038 mm mesh size) in 0.5 m deep was randomly assigned. Soil respiration and its components were measured using an LI-6400 portable photosynthesis system attached to a soil CO_2 flux chamber (800 cm^3 in total volume; LI-COR 6400-09 TC, LI-COR Inc., Lincoln, NE, USA). The CO_2 efflux measured in the root-free plots reflects only microbial respiration, while CO_2 efflux measured in the whole-soil plots (roots are not removed) resulted from both microbial and root respiration. The difference between the CO_2 efflux values for root-free plots and whole-soil plots was used to indicate root respiration. However, we observed that the soil temperature and moisture in root-free plots were significantly higher than those in whole-soil plots. The actual root respiration would be underestimated if it is directly calculated from the difference of measured CO_2 flux between the whole-soil plot and root-free plot. To eliminate this error, we corrected the measured microbial respiration by using the linear Eq. (5), simulating the relationship between microbial respiration, soil temperature and soil moisture in root-free plots:

$$MR_{measured} = a \times T + b \times W + c \quad (5)$$

where $MR_{measured}$, T and W are the microbial respiration (μmol CO_2 m^{-2} s^{-1}), soil temperature (°C) and volumetric soil water content (%) measured in the root-free plot, respectively. a, b and c are coefficients relevant to soil temperature and moisture.

Then, we determined the corrected microbial respiration ($MR_{corrected}$) using the soil temperature and moisture in the whole-soil plot. Root respiration (RR) calculated by the difference between the SR and the $MR_{corrected}$ is as follows:

$$RR = SR - MR_{corrected} \quad (6)$$

Soil temperature at the depth of 5 cm was determined using a thermocouple probe connected to the LI-6400 adjacent to each PVC collar, and volumetric soil content in the 0–10 cm soil layers was measured using a TRIME TDR probe (IMKO, Ettlingen, Germany) adjacent to the same sites after soil temperature measurements. The root length production was measured using the minirhizotrons technique [44]. Peak aboveground biomass (AGB) was estimated by harvesting plant tissues above the soil surface from one 0.5 × 0.5 m quadrats at each plot in late September of both years. After aboveground plant residues cleaned, soil samples to depths of 10 cm were collected. Roots were collected from soil samples to determine belowground biomass (BGB). WSOC was measured using an automated total organic C analyzer

(TOC-Vcph, Shimadzu, Japan) [45]. SMBC was determined using the chloroform fumigation extraction method [46].

2.2.5. Biomass data collection

Field harvest was conducted in mid or late August each year from 1982 to 2011, when the standing biomass reached its maximum. For each harvest in each year, 15 quadrats (1 × 1 m) were selected along a transect (300 × 100 m). Aboveground biomass was clipped and dried at 65°C to constant weight. Between 1982 and 1992, the degraded grassland recovered rapidly and biomass production increased almost linearly. It was mainly caused by the exclusion of human disturbance, particularly overgrazing. After 1992, grasslands assumed a relatively balanced state with lower variation in productivity and diversity. Further variation in productivity was likely caused primarily by climatic variation. We therefore used the peak aboveground biomass during 1992–2011 to evaluate the impacts of climate variability on grassland productivity. Mean daily temperature and precipitation during 1992–2011 were obtained from a weather station established in 1982, located only 0.9 km from the surveyed transect.

2.3. Data analyses

2.3.1. One-way analysis of variance

A one-way analysis of variance (ANOVA) followed by Tukey's HSD test was conducted to determine the effect of grazing exclusion time on grassland root traits (*RB, RLD, RSA*, plant functional group richness, plant species richness), the differences of root traits (*SRL, SRS*) and proportion in root community between plant functional groups, and the effects of clipping over time on soil respiration, microbial respiration, root respiration, soil temperature and soil moisture. Differences were considered significant for all statistical tests at $P < 0.05$. All the statistical analyses were conducted using IBM SPSS 18.0 (IBM, USA). Graphs were created with Sigma plot 12.5 (Systat Software, USA).

2.3.2. Partial least squares

Partial least squares (PLS) regression was used to analyze the responses of grassland productivity to variation in daily temperature and precipitation during all 365 days of the year based on data for 1992–2011. The two major outputs of PLS analysis are the variable importance in the projection (VIP) and standardized model coefficients. The VIP values reflect the importance of all independent variables for explaining variation in dependent variables. The VIP threshold for considering variables as important is often set to 0.8. The standardized model coefficients indicate the strength and direction of the impacts of each variable in the PLS model. The root-mean-square errors (RMSEs) of the regression analyses were calculated to determine the accuracy of the PLS model. In the PLS analyses, periods with VIP greater than 0.8 and high absolute values of model coefficients represent the relevant phases influencing grassland productivity. Positive model coefficients indicate that increasing temperature or precipitation during the respective period should increase ANPP, while negative model coefficients imply negative impacts on productivity.

3. Results and discussion

3.1. Effects of grazing exclusion on grassland root biomass and morphological traits

Results demonstrated that long-term grazing exclusion significantly increased grassland root biomass, root length density and root surface area ($P < 0.05$) (**Table 1**). The improved root biomass was mainly due to the increased aboveground productivity driven by the compensatory growth of dominant plant species after grazing removal [11, 18]. In the absence of herbivores, plants produced more roots to explore soil resource for aboveground growth, inducing increases in grassland total root length and surface area [47, 48]. Besides, our results indicated that the response of plant belowground richness to grazing exclusion followed a hump-like pattern, similar with responses of plant aboveground richness and diversity to grazing exclusion [11, 16], but with an earlier peak in the early-restoration stage (site GE05). Possibly long-term grazing exclusion caused a drastic decrease in bud bank size of forbs, followed with the decline or even disappearance of plant species relying on resprouting from bud bank after disturbance [49].

3.2. Root traits and proportional changes of five plant functional groups after grazing exclusion

Plant *SRL* and *SRS* showed significant differences between five plant functional groups ($P < 0.05$). Grasses had a much higher specific root length and specific root surface area than forbs. In detail, for *SRL*, PB had the highest value of 11.80 m g^{-1}, tripling that of SS (3.46 m g^{-1}), while PR and AB had similar *SRL* value, higher than that of PF (**Figure 1a**). For *SRS*, there were no marked variations among PR, PF and AB, and they were significantly higher and lower than those of SS and PB, respectively (**Figure 1b**). Our results indicated that plant functional groups differed significantly in their proportions ($P < 0.05$) (**Figure 2a–d**). As the predominant plant functional groups, PB and PF accounted for more than 50% in total. Based on root biomass, proportions of PR and PB significantly increased with a significant decrease in PF proportion after long-term grazing exclusion ($P < 0.05$), and SS and AB showed little change ($P > 0.05$) (**Figure 2a**). Based on root length density and root surface area, grazing exclusion significantly increased PR proportion and decreased PF proportion ($P < 0.05$), while PB and SS show little responses to grazing exclusion ($P > 0.05$) (**Figure 2b**, **c**). Interestingly, with the prolonged grazing exclusion years,

Site	Root biomass (g m^{-2})	*RLD* (10^3 m m^{-3})	*RSA* (10^4 cm^2 m^{-3})	PFG Richness	Plant species richness
GG	163.32 ± 11.27 c	1.76 ± 0.19 c	3.04 ± 0.32 b	3.55 ± 0.29	10.78 ± 0.74ab
GE05	172.22 ± 17.07 bc	2.10 ± 0.22 bc	3.37 ± 0.29 b	4.33 ± 0.17	13.00 ± 0.62 a
GE09	170.45 ± 12.35 bc	2.10 ± 0.32 bc	3.43 ± 0.30 b	4.00 ± 0.24	8.56 ± 0.76 bc
GE22	236.61 ± 21.52 ab	3.03 ± 0.29 ab	4.02 ± 0.24 b	3.44 ± 0.24	6.78 ± 0.74 c
GE30	244.41 ± 21.25 a	3.85 ± 0.25 a	5.21 ± 0.30 a	3.56 ± 0.24	7.44 ± 0.90 c

Different lowercase letters indicate significant differences ($P < 0.05$) between five study sites.

Table 1. Root biomass, root length density (*RLD*), root surface areas (*RSA*), plant functional group (PFG) richness and plant species richness in study sites.

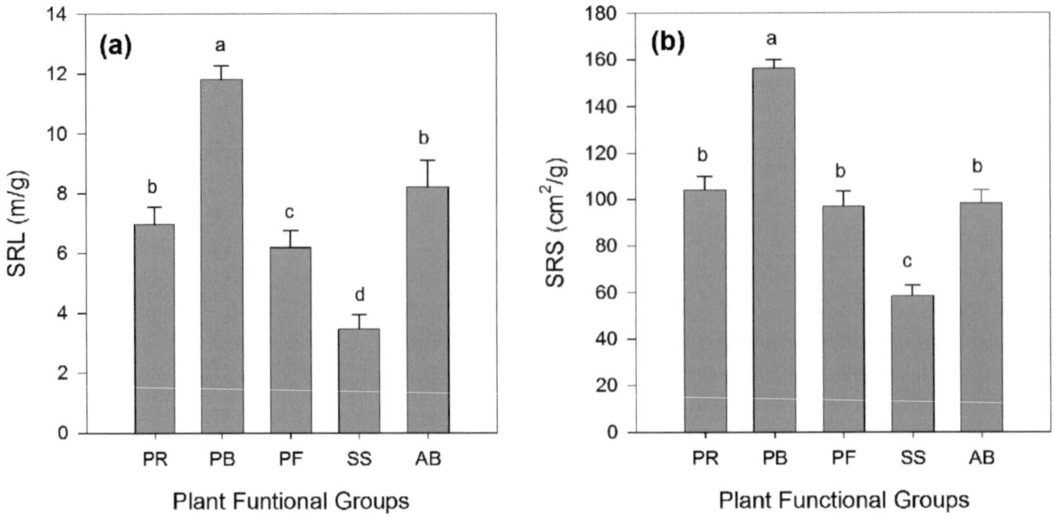

Figure 1. *SRL* traits (a) and *SRS* traits (b) of five plant functional groups. Different lowercase letters indicate significant differences ($P < 0.05$) between plant functional groups.

proportions of PR and PB in plant species richness significantly increased ($P < 0.05$), and those of PF and AB significantly decreased ($P < 0.05$), while SS showed little fluctuation ($P > 0.05$) (**Figure 2d**).

As the guerrilla plant species, PR had advantages in spatial propagation and exploration of adjacent nutrient patches by increasing rhizome and root length after grazing exclusion [50]. Additionally, dispersal by rhizomes allowed temporal release of PR plants from their natural enemies (i.e., root herbivores and pathogens), which stimulated plant growth in return [51]. The compositional changes of plant functional groups mainly resulted from their different responses to improved soil resources after grazing exclusion [52]. Compared with forbs, grasses had a stronger correlation with soil N [16], and grasses' higher *SRL* and *SRS* consolidated their superiority in acquiring soil resources [20]. Given that nitrogen deposition often occurs with accompanying rainfall events, which forms water and nutrient pulses [53], plants with larger root systems (i.e., grasses) gained more benefit than smaller plants at the start of the nutrient pulse [54]. Therefore, our study indicated that the hierarchy of root system size and root traits among five plant functional groups determined grassland root pattern in semiarid grassland after long-term grazing exclusion.

3.3. Effect of clipping on soil respiration

Clipping significantly reduced the mean soil respiration by 14.7% ($P < 0.001$) and 11.4% ($P < 0.05$) in 2014 and 2015, respectively (**Table 2, Figure 3a**). Previous research has reported that clipping could decrease the soil respiration in grassland ecosystems, which was most likely due to the restriction of translocation of photosynthate from aboveground plant tissues to roots and rhizosphere microorganisms [31]. In addition, clipping increased soil temperature by 0.6°C ($P > 0.05$)

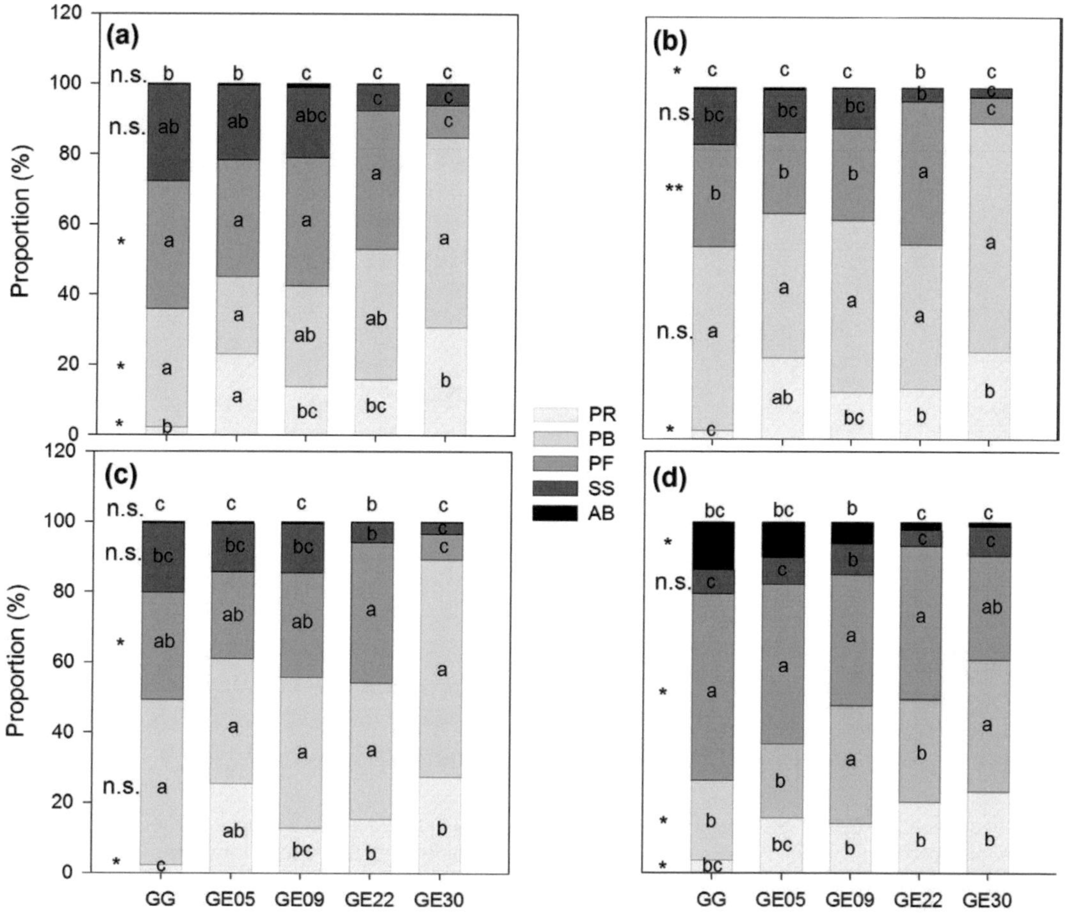

Figure 2. Distribution proportions of five plant functional groups in root biomass (a), root length density (b), root surface area (c) and plant species richness (d) in grazing grassland (GG), grassland with grazing exclusion for 5 years (GE05), 9 years (GE09), 22 years (GE22) and 30 years (GE30), respectively. Different lowercase letters indicate significant differences ($P < 0.05$) between five plant functional groups for each grassland type, and n.s. indicates no significant difference ($P > 0.05$) for each plant functional group between five grassland types; * and ** indicate significant differences for each plant functional groups between five grasslands in $P < 0.05$ level and $P < 0.01$ level, respectively.

Source of variation	*SR*		*MR*		*RR*		ST		SM	
	2014	2015	2014	2015	2014	2015	2014	2015	2014	2015
Treatment	<0.001	0.034	0.005	0.008	<0.001	0.092	0.110	0.023	0.305	0.544
Time	<0.001	<0.001	<0.001	<0.001	<0.001	<0.001	<0.001	<0.001	<0.001	<0.001
Treatment × time	<0.001	0.011	0.746	0.134	<0.001	0.078	0.289	0.059	0.742	0.286

Table 2. *P*-values of repeated measures ANOVA of total soil respiration (*SR*), microbial respiration (*MR*), root respiration (*RR*), soil temperature (ST) and soil moisture (SM) in a temperate grassland of Loess Plateau.

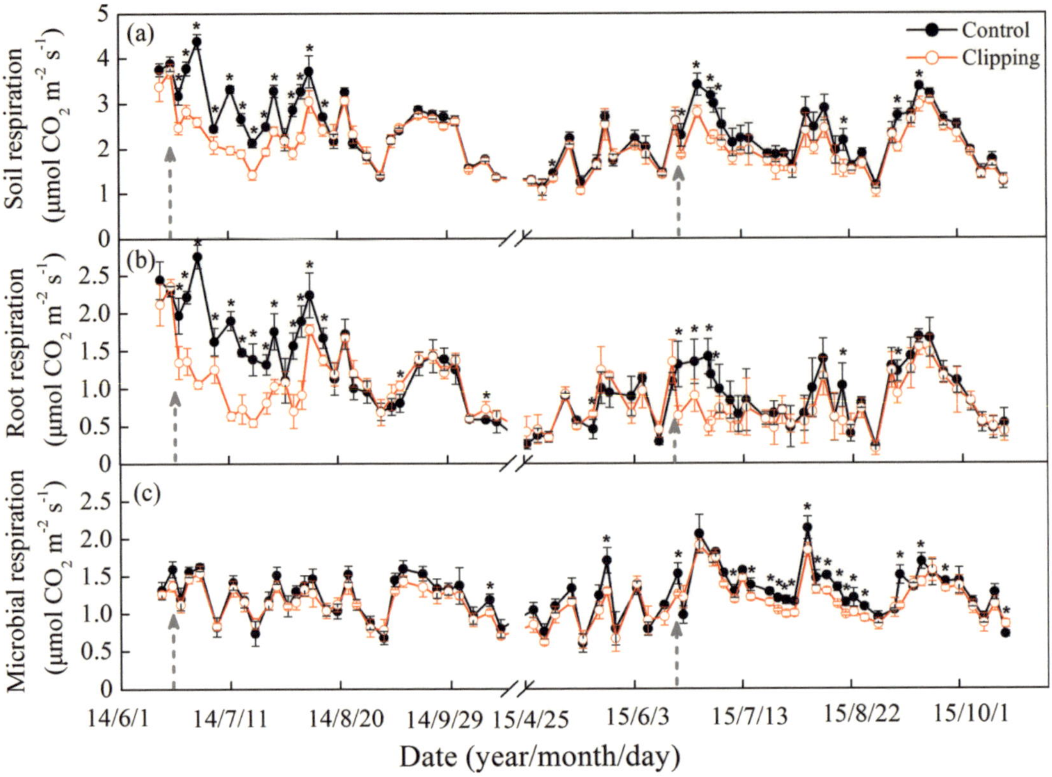

Figure 3. Seasonal variations of soil respiration and its components in the control and clipping treatments. Values are means, standard deviations ($n = 5$). Asterisks denote significant difference ($P < 0.05$) between treatments. Arrows indicate clipping dates.

in 2014 and 1.3°C ($P < 0.05$) in 2015 in our study (**Figure 4**). We speculated that there was a potential increase in soil respiration driven by soil temperature, because higher soil temperature has been reported to stimulate the activities of plant roots and soil microbes [29]. However, the increase of soil respiration due to elevated soil temperature may not compensate for decrease in soil respiration caused by reduced photosynthesis, leading to the decrease in soil respiration after clipping.

3.4. Effect of clipping on root respiration

In the present study, clipping reduced the mean root respiration by 22.1% ($P < 0.001$) and 13.3% ($P > 0.05$) in 2014 and 2015, respectively (**Table 2**, **Figure 3b**). We found a prompt response in root respiration in the first measurements after two days of clipping treatment, following the sharp reduction of 49.2 and 26.4% within two weeks after treatment in 2014 and 2015, respectively (**Figure 3b**). We also found that the sharp decrease in root respiration was consistent with the sudden reduction of root production in the same periods (**Figures 3b** and **5a**). Considering the significant correlation between the root production and root respiration (**Figure 5a**), we attributed the decrease of root respiration after clipping to the limited supply

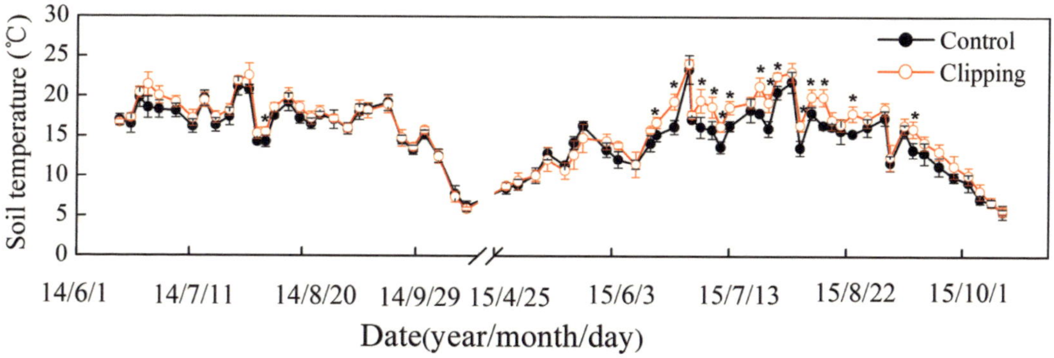

Figure 4. Seasonal variations of soil temperature in the control and clipping treatments. Values are means, standard deviations ($n = 5$).

of substrate for root growth and production. However, in September–October in 2014 and April–May in 2015, a higher root respiration was observed in the clipping plots (**Figure 3b**). Previous studies by Wan et al. [55] and Zhou et al. [33] reported that clipping could stimulate root respiration by promoting plant regrowth and root biomass. In our study, the higher root production observed in clipping plots in September–October in 2014 and April–May in 2015 might be responsible for the higher root respiration in the same periods (**Figures 3c** and **5a**).

3.5. Effect of clipping on microbial respiration

Microbial respiration exhibited relatively constant lower values in clipping plots almost throughout the study period in our study. Clipping significantly reduced microbial respiration by 6.0% ($P < 0.05$) and 9.9% ($P < 0.05$) in 2014 and 2015, respectively (**Table 2**, **Figure 3c**). The main explanation of this result was the reduced supply of labile C for mineralization by soil microorganism after clipping [56]. In the present study, clipping reduced the WSOC by 20.6% ($P > 0.05$) and 27.1% ($P < 0.05$) in 2014 and 2015, respectively (**Figure 5b**). The decrease of WSOC might be responsible for the reduction of SMBC in clipping plots in our study (**Figure 5c**), because WSOC was one of the main labile C substrates for soil microorganism. In addition, SMBC was reported to be significantly related to microbial respiration in previous research [57], which was similar to our results (R^2=0.88, $P < 0.05$). Hence, we attributed the decrease of microbial respiration after clipping to the reduction of available C supply for microbial mineralization.

3.6. Response of grassland productivity to variation in daily temperature

Between 1992 and 2011, the average harvest date of peak aboveground biomass for grassland at Yunwushan National Nature Reserve was 15th of August. The 365 daily temperature values between the previous September and August of the year of harvest were used as independent variables in the PLS regression. A low root-mean-square error (RMSE) of 8.13 g m^{-2} for the resulting PLS model indicated that the model was a good fit for the data. Based on the VIP and standardized model coefficients of the PLS analysis, we found that warming during different periods had varied impacts on grassland productivity (**Figure 6**).

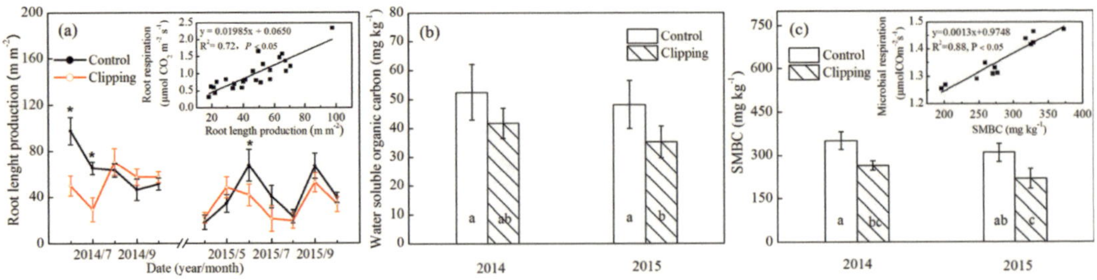

Figure 5. Comparison of root length production (a), water-soluble organic carbon in trenched plots (b) and soil microbial biomass C in trenched plots (c) among treatments. Asterisks and different letters denote significant difference ($P < 0.05$) between treatments. Arrows indicate clipping dates.

Between 30 March and 30 May, model coefficients for temperature analysis (**Figure 6**) were always positive and VIP values mostly exceeded 0.8 (the threshold for variable importance), indicating that warming in April and May increases grassland productivity. During 31 May–1 August, model coefficients were consistently negative and VIP values were mostly important, implying that temperature increase in summer (June–July) depressed productivity, forming a striking contrast with the impacts of spring warming. It was of interest that the relevant periods influencing productivity, as identified by PLS regression, were almost the same as the phases of plant growth (i.e., the early and middle stages of the growing season) at our study area. No obvious impacts of temperature variation in August on grassland productivity were apparent. During September–October (the senescence period for vegetation), most model coefficients were negative, indicating that high temperature at that time was unfavorable for productivity of the following year. During 1 November–29 March, the dormancy period, model coefficients were mostly negative, although this phase also included some short intervals with positive coefficients. This variation might indicate that dormancy for grassland is a complex physiological and ecological process. Moreover, it seems possible that the strength of temperature impacts varies throughout the dormancy period. Taking a broader view at model coefficients and aiming at consistency with established phonological phases, we interpreted the entire period (November–March) as another relevant period during which temperature increases appeared to reduce grassland productivity.

3.7. Response of grassland productivity to variation in daily precipitation

The 365 daily precipitation values between the previous September and August were also used as independent variables in the PLS analysis. The resulting model still proved to be a good fit for the data, with an RMSE of 6.53 g m^{-2}. In contrast to the positive effects of higher precipitation in June and July, increasing rainfall during the senescence period (September–October) and the early growing season (April–May) was correlated with low productivity (**Figure 6**). Similar to temperature effects in August, no significant relationship was found between grassland ANPP and precipitation in August. During the dormancy period, there was no consistent correlation between precipitation and productivity. Positive impacts were almost offset by negative ones.

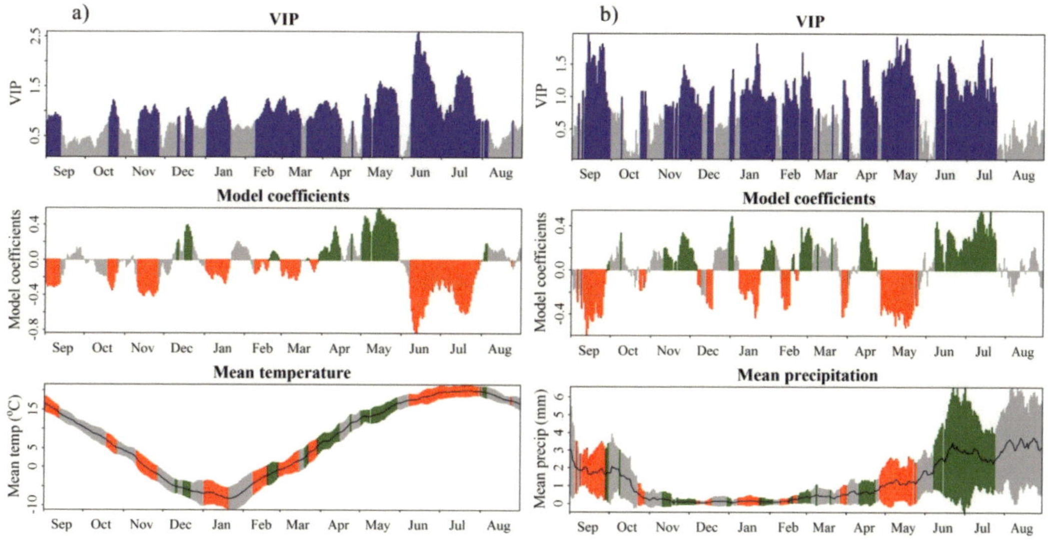

Figure 6. Results of partial least squares (PLS) regression correlating grassland productivity at Yunwushan during 1992–2011 with 15-day running means of (a) daily mean temperature and (b) daily precipitation from previous September to August. Blue bars in the top row indicate that VIP values are greater than 0.8, the threshold for variable importance. In the middle row, red color means model coefficients are negative and important, while green color indicates important positive relationships between grassland productivity and climate variables. The black lines in the bottom panel stand for daily mean temperature and precipitation, while gray, green and red areas represent the standard deviation of daily climate variables.

The increased temperature with reduced precipitation in spring (April–May) could improve grassland productivity. Biomass produced in spring is often believed to be limited by cold temperatures at mid or high latitude [58]. Temperature increases early in the growing season may stimulate plant growth directly by raising leaf temperatures or indirectly by increasing water absorption and N mineralization (**Figure 7**) [40]. Additionally, warmer springs also likely accelerate snowmelt and advance spring greening [59], which might lengthen the growing season and result in increased photosynthesis and carbon acquisition [60]. In contrast to some studies reporting that more precipitation during April–May promoted grassland productivity [39], we found a negative relationship between these variables. To some extent, this discrepancy can be explained by the site hydrology. Frequent winter snow (lasting from November to March) in our study area provides sufficient soil water for plant growth in early spring. The sporadic precipitation during April–May (with an average of 59.5 mm during these two months between 1992 and 2011) may not have important direct impacts on productivity. In contrast, low air and soil temperature, as well as limited solar radiation caused by frequent rain events in May, might partially explain the negative correlations between spring rainfall and grassland productivity.

Warming in summer coinciding with drought can generate physiological stress for plant growth (**Figure 7**) [61], which can explain the reduced productivity in our study area. Moreover, increases in summer temperature can also lower ANPP, perhaps by reducing soil moisture through increased evapotranspiration. Decrease in precipitation amounts and lengthening of intervals between precipitations events during the past 20 years further reduced soil water

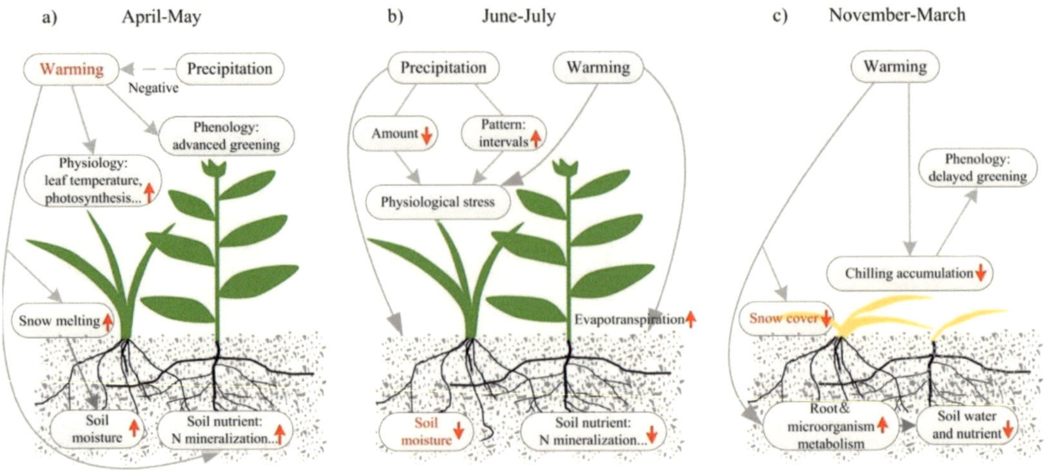

Figure 7. Potential relationships between grassland productivity and climate variability during (a) April–May, (b) June–July and (c) November–March at Yunwushan.

availability in our study region. This is in line with the hypothesis that impacts of climate variation and change on plant productivity might occur via variability in soil moisture [36]. Continuous warming and drought in summer could also affect N mineralization negatively and limit soil resource availability, thereby reducing productivity.

PLS regression did not detect a response of grassland productivity to climatic variation in August. Compared to climate variation during June–July, August shows more variable temperature and precipitation in our study region, although August is cooler on average than July. For instance, the coefficient of variation (CV) of precipitation in August between 1992 and 2011 was 53.3%, while it was only 33.5% for June–July. It is also worth noting, however, that in our study biomass was mostly harvested around the 15th of August, so that the vegetation was only exposed to half a month of August conditions.

Increases in temperature and precipitation during September–October in the previous year were negatively correlated with productivity in the current year, which can be partially explained by the widely reported delays of senescence caused by warming and wetting later in the year [62]. Delay in the senescence period may be related to some extent to increased soil nutrient and water depletion. This would imply that fewer resources may have been available for biomass production in the following year.

While some studies reported that weather during the dormancy period had limited impacts on grassland productivity [38], such effects may become more important, as temperature in winter further increases. Our results indicated that high temperatures during the dormancy period were negatively correlated with productivity. This is consistent with warming experiments in two limestone grasslands in the UK, which showed that winter heating combined with drought reduced the biomass of both communities [63]. Warmer winter can lead to some unanticipated consequences (**Figure 7**). The most direct impacts have been a shortening of the snow season and a reduction in snow cover, which have been observed in our study area. Declines in the area and depth of snow cover may expose the land surface to more frequent freezing events, exerting negative effects on plant growth. This is supported by observations in northern Scandinavia where extensive areas of vegetation died due to loss of snow cover after extreme winter warming in December 2007 [64]. Increased demands of soil nutrients and water due to accelerated root and microorganism metabolism caused by winter warming might also contribute to the productivity reduction. Finally, variation in spring phenology can also help explain this phenomenon. The timing of spring phenology in most temperate plants results from the interplay of winter cold and spring heat. Plants that evolved in temperate climates fall dormant in autumn to protect themselves from winter freezes and will only resume growth in spring when they have been exposed sufficiently to cold conditions [65]. Temperature increases in spring can advance spring phenology (e.g., greening for grassland), but warming in winter may delay the fulfillment of chilling requirements and thus lead to a slowdown in the advance of spring events or even later onset of spring phenology [65, 66]. The advancing trend in spring greening still dominates climate change responses of plants in our study region so far, since chilling requirements for vegetation are easily satisfied in all winters under the present cold climate with a mean temperature of −2.6°C for the dormancy period. As global warming progresses, especially when rates and effects of warming in winter exceed those in spring, advances in greening might be slowed or even turn into delays. We therefore recommend increased scientific attention to impacts of winter warming on grassland productivity and the timing of spring phenology events.

4. Conclusion

Based on the results of the long-term experiments highlighted in this chapter, grassland root biomass and root morphological traits significantly increased after long-term grazing exclusion, accompanying with significantly declined plant species richness. The higher *SRL* and *SRS* may determine the increased proportion of grasses. The root respiration and microbial respiration exhibited different response patterns to the spring clipping. Compared with the relatively constant lower values in clipping plots almost throughout the study period for microbial respiration, root respiration fluctuated greatly in response to clipping treatment. In addition, soil water content could affect the response of soil respiration and its components to clipping in aspect of magnitudes and resilience in the semiarid grassland ecosystem. PLS regression between ANPP and daily climate variables during the past 20 years successfully delineated how timing of temperature and precipitation variability affected grassland productivity on the Loess Plateau in China. The analysis of productivity responses should account not only for the magnitude of climate variation but also for its timing.

5. Future perspective

The land use practices have substantially improved the above- and belowground ecophysiological processes in the degraded semiarid grasslands. At the site level, plant roots and soil organisms are two important parts in the plant-soil feedback system, which affect the growth of plants and formation of interspecific competition relationship and vegetation pattern. Therefore, future studies will focus on evaluating the rules of plant-soil feedback during the plant growth process. At the macro level, has climate variation contributed to grassland restoration, and how the LUCC and human activities have affected grasslands in the ecological projects implemented regions? These are still open questions that need to be addressed. These studies will provide insights in effective management measurements for the ecological restoration projects and may serve as guidelines for government and policy makers in adjusting future ecological policies and managing grassland production in the western China for adapting to climate conditions.

Acknowledgements

This work was supported by National Key Research Program of China (2016YFC0500700), National Natural Science Foundation of China (31602004, 41230852, 31601987, 41601586, 41601053), Deployment Program of the Chinese Academy of Sciences (KJZD-EW-TZ-G10) and Technical System of Modern Forage Industry of the Ministry of Agriculture (CARS-35-40). We thank Weijun Li, Xiuyun Ji and the management bureau of Yunwushan Grassland Natural Reserve for their help with the field experiments.

Author details

Jimin Cheng[1,2]*, Chengcheng Gang[1,2], Liang Guo[1,2], Wei Li[1,2], Jingwei Jin[1,2], Jishuai Su[3] and Lin Wei[1,2]

*Address all correspondence to: gyzcjm@ms.iswc.ac.cn

1 Institute of Soil and Water Conservation, Northwest A&F University, Yangling, China

2 Institute of Soil and Water Conservation, Chinese Academy of Sciences and Ministry of Water Resources, Yangling, China

3 College of Animal Science and Technology, Northwest A&F University, Yangling, China

References

[1] O'Mara FP. The role of grasslands in food security and climate change. Annals of Botany. 2012;**110**:1263–1270. DOI: 10.1093/aob/mcs209

[2] Scurlock JMO, Hall DO. The global carbon sink: A grassland perspective. Global Change Biology. 1998;**4**:229–233. DOI: 10.1046/j.1365-2486.1998.00151.x

[3] Gang CC, Zhou W, Wang ZQ, Chen YZ, Li JL, Chen JM, et al. Comparative assessment of grassland NPP dynamics in response to climate change in China, North America, Europe and Australia from 1981 to 2010. Journal of Agronomy and Crop Science. 2015;**201**:57–68. DOI: 10.1111/jac.12088

[4] Zhou W, Gang C, Zhou L, Chen Y, Li J, Ju W, Odeh I. Dynamic of grassland vegetation degradation and its quantitative assessment in the northwest China. Acta Oecologica. 2014;**55**:86–96. DOI: 10.1016/j.actao.2013.12.006

[5] Ni J. Carbon storage in grasslands of China. Journal of Arid Environments. 2002;**50**:205–218. DOI: 10.1006/jare.2201.0902

[6] Mu SJ, Yang HF, Li JL, Chen YZ, Gang CC, Zhou W, et al. Spatio-temporal dynamics of vegetation coverage and its relationship with climate factors in Inner Mongolia, China. Journal of Geographical Sciences. 2013;**23**:231–246. DOI: 10.1007/s11442-013-1006-x

[7] Kang L, Han X, Zhang Z, Sun OJ. Grassland ecosystems in China: Review of current knowledge and research advancement. Philosophical Transactions of the Royal Society B: Biological Sciences. 2007;**362**:997–1008. DOI: 10.1098/rstb.2007.2029

[8] Wei L, Liu J, Su J, Jing G, Zhao J, Cheng J, Jin J. Effect of clipping on soil respiration components in temperate grassland of Loess Plateau. European Journal of Soil Biology. 2016;**75**:157–167. DOI: 10.1016/j.ejsobi.2016.06.003

[9] Akiyama T, Kawamura K. Grassland degradation in China: Methods of monitoring, management and restoration. Grassland Science. 2007;**53**:1–17. DOI: 10.1111/j.1744-697X.2007.00073.x

[10] Jiang GM, Han XG, Wu JG. Restoration and management of the Inner Mongolia grassland require a sustainable strategy. AMBIO. 2006;**35**:269–270. DOI: 10.1579/06-S-158.1

[11] Wu GL, Du GZ, Liu ZH, Thirgood S. Effect of fencing and grazing on a Kobresia-dominated meadow in the Qinghai-Tibetan Plateau. Plant and Soil. 2009;**319**:115–126. DOI: 10.1007/s11104-008-9854-3

[12] Deng L, Zhang ZN, Shangguan ZP. Long-term fencing effects on plant diversity and soil properties in China. Soil & Tillage Research. 2014;**137**:7–15. DOI: 10.1016/j.still.2013.11.002

[13] Shrestha G, Stahl PD. Carbon accumulation and storage in semi-arid sagebrush steppe: Effects of long-term grazing exclusion. Agriculture Ecosystems & Environment. 2008;**125**:173–181. DOI: 10.1016/j.agee.2007.12.007

[14] Golodets C, Kigel J, Sternberg M. Recovery of plant species composition and ecosystem function after cessation of grazing in a Mediterranean grassland. Plant and Soil. 2010;**329**:365–378. DOI: 10.1007/s11104-009-0164-1

[15] Jing ZB, Cheng JM, Su JS, Bai Y, Jin JW. Changes in plant community composition and soil properties under 3-decade grazing exclusion in semiarid grassland. Ecological Engineering. 2014;**64**:171–178. DOI: 10.1016/j.ecoleng.2013.12.023

[16] Qiu LP, Wei XR, Zhang XC, Cheng JM. Ecosystem carbon and nitrogen accumulation after grazing exclusion in semiarid grassland. PloS One. 2013;**8**:e55433. DOI: 10.1371/journal.pone.0055433

[17] Jing ZB, Cheng JM, Jin JW, Su JS, Bai Y. Revegetation as an efficient means of improving the diversity and abundance of soil eukaryotes in the Loess Plateau of China. Ecological Engineering. 2014;**70**:169–174. DOI: 10.1016/j.ecoleng.2014.05.011

[18] Liu M, Liu GH, Wu X, Wang H, Che L. Vegetation traits and soil properties in response to utilization patterns of grassland in Hulun Buir City, Inner Mongolia, China. Chinese Geographical Science. 2014;**24**:471–478. DOI: 10.1007/s11769-014-0706-1

[19] Jackson RB, Caldwell MM. Geostatistical patterns of soil heterogeneity around individual perennial plants. Journal of Ecology. 1993;**81**:683–692. DOI: 10.2307/2261666

[20] Li HB, Ma QH, Li HG, Zhang FS, Rengel Z, Shen JB. Root morphological responses to localized nutrient supply differ among crop species with contrasting root traits. Plant and Soil. 2014;**376**:151–163. DOI: 10.1007/s11104-013-1965-9

[21] Pohl M, Stroude R, Buttler A, Rixen C. Functional traits and root morphology of alpine plants. Annals of Botany. 2011;**108**:537–545. DOI: 10.1093/aob/mcr169

[22] Johnson HA, Biondini ME. Root morphological plasticity and nitrogen uptake of 59 plant species from the Great Plains grasslands, USA. Basic and Applied Ecology. 2001;**2**:127–143. DOI: 10.1078/1439-1791-00044

[23] Fort F, Cruz P, Jouany C. Hierarchy of root functional trait values and plasticity drive early-stage competition for water and phosphorus among grasses. Functional Ecology. 2014;**28**:1030–1040. DOI: 10.1111/1365-2435.12217

[24] Luo YQ, Wan SQ, Hui DF, Wallace LL. Acclimatization of soil respiration to warming in a tall grass prairie. Nature. 2001;**413:**622–625. DOI: 10.1038/35098065

[25] Raich JW, Schlesinger WH. The global carbon dioxide flux in soil respiration and its relationship to vegetation and climate. Tellus B. 1992;**44:**81–99. DOI: 10.1034/j.1600-0889.1992.t01-1-00001.x

[26] Han Y, Zhang Z, Wang CH, Jiang FH, Xia JY. Effects of mowing and nitrogen addition on soil respiration in three patches in an oldfield grassland in Inner Mongolia. Journal of Plant Ecology. 2012;**5:**219–228. DOI: 10.1093/jpe/rtr015

[27] Bahn M, Knapp M, Garajova Z, Pfahringer N, Cernusca A. Root respiration in temperate mountain grasslands differing in land use. Global Change Biology. 2006;**12:**995–1006. DOI: 10.1111/j.1365-2486.2006.01144.x

[28] Cao GM, Tang YH, Mo WH, Wang YS, Li YN, et al. Grazing intensity alters soil respiration in an alpine meadow on the Tibetan plateau. Soil Biology and Biochemistry. 2004;**36:**237–243. DOI: 10.1016/j.soilbio.2003.09.010

[29] Jia XX, Shao MA, Wei XR. Responses of soil respiration to N addition, burning and clipping in temperate semiarid grassland in northern China. Agricultural and Forest Meteorology. 2012;**166**:32–40. DOI: 10.1016/j.agrformet.2012.05.022

[30] Ryan MG, Law BE. Interpreting, measuring, and modeling soil respiration. Biogeochemistry. 2005;**73**:3–27. DOI: 10.1007/s10533-004-5167-7

[31] Wan SQ, Luo YQ. Substrate regulation of soil respiration in a tallgrass prairie: Results of a clipping and shading experiment. Global Biogeochemical Cycles. 2003;**17**:1054. DOI: 10.1029/2002GB001971

[32] Schindlbacher A, Zechmeister-Boltenstern S, Jandl R. Carbon losses due to soil warming: Do autotrophic and heterotrophic soil respiration respond equally? Global Change Biology. 2009;**15**:901–913. DOI: 10.1111/j.1365-2486.2008.01757.x

[33] Zhou XH, Wan SQ, Luo YQ. Source components and interannual variability of soil CO_2 efflux under experimental warming and clipping in a grassland ecosystem. Global Change Biology. 2007;**13**:761–775. DOI: 10.1111/j.1365-2486.2007.01333.x

[34] Knapp AK, Smith MD. Variation among biomes in temporal dynamics of aboveground primary production. Science. 2001;**291**:481–484. DOI: 10.1126/science.291.5503.481

[35] Lauenroth WK, Sala OE. Long-term forage production of North American shortgrass steppe. Ecological Application. 1992;**2**:397–403. DOI: 10.2307/1941874

[36] Oesterheld M, Loreti J, Semmartin M, Sala OE. Inter-annual variation in primary production of a semi-arid grassland related to previous-year production. Journal of Vegetation Science. 2001;**12**:137–142. DOI: 10.1111/j.1654-1103.2001.tb02624.x

[37] Hsu JS, Adler PB. Anticipating changes in variability of grassland production due to increases in interannual precipitation variability. Ecosphere. 2014;**5**:1–15. DOI: 10.1890/ES13-00210.1

[38] La Pierre KJ, Yuan SH, Chang CC, Avolio ML, Hallett LM, Schreck T, et al. Explaining temporal variation in above-ground productivity in a mesic grassland: The role of climate and flowering. Journal of Ecology. 2011;**99**:1250–1262. DOI: 10.1111/j.1365-2745.2011.01844.x

[39] Robinson TMP, La Pierre KJ, Vadeboncoeur MA, Byrne KM, Thomey ML, Colby SE. Seasonal, not annual precipitation drives community productivity across ecosystems. Oikos. 2013;**122**:727–738. DOI: 10.1111/j.1600-0706.2012.20655.x

[40] Craine JM, Nippert JB, Elmore AJ, et al. Timing of climate variability and grassland productivity. Proceedings of the National Academy of Sciences. 2012;**109**:3401–3405. DOI: 10.1073/pnas.1118438109

[41] Bai YF, Han XG, Wu JG, Chen ZZ, Li LH. Ecosystem stability and compensatory effects in the Inner Mongolia grassland. Nature. 2004;**431**:181–184. DOI: 10.1038/nature02850

[42] Fransen B, Blijjenberg J, de Kroon H. Root morphological and physiological plasticity of perennial grass species and the exploitation of spatial and temporal heterogeneous nutrient patches. Plant and Soil. 1999;**211**:179–189. DOI: 10.1007/s004420050527

[43] Moyano FE, Kutsch WL, Schulze ED. Response of mycorrhizal, rhizosphere and soil basal respiration to temperature and photosynthesis in a barley field. Soil Biology and Biochemistry. 2007;**39:**843–853. DOI: 10.1016/j.soilbio.2006.10.001

[44] Bai WM, Wan S, Niu S, Liu W, Chen Q, Wang Q, et al. Increased temperature and precipitation interact to affect root production, mortality, and turnover in a temperate steppe: Implications for ecosystem C cycling. Global Change Biology. 2010;**16:**1306–1316. DOI: 10.1111/j.1365-2486.2009.02019.x

[45] Liang BC, Mackenzie AF, Schnitzer M. Management-induced change in labile soil organic matter under continuous corn in eastern Canadian soils. Biology and Fertility of Soils. 1998;**26:**88–94. DOI: 10.1007/s003740050348

[46] Vance ED, Brookes PC, Jenkinson DS. An extraction method for measuring soil microbial biomass C. Soil Biology and Biochemistry. 1987;**19:**703–707. DOI: 10.1016/0038-0717(87)90052-6

[47] Bezemer TM, Van der Putten WH. Ecology: Diversity and stability in plant communities. Nature. 2007;**446**:E6–E7. DOI: 10.1038/nature05749

[48] Kuka K, Illerhaus B, Fox CA, Joschko M. X-ray computed microtomography for the study of the soil–root relationship in grassland soils. Vadose Zone Journal. 2013;**12**:1–10. DOI: 10.2136/vzj2013.01.0014

[49] Fidelis A, Appezzato-da-Glória B, Pillar VD, Pfadenhauer J. Does disturbance affect bud bank size and belowground structures diversity in Brazilian subtropical grasslands? Flora. 2014;**209**:110–116. DOI: 10.1016/j.flora.2013.12.003

[50] Amiaud B, Touzard B, Bonis A, Bouzillé J-B. After grazing exclusion, is there any modification of strategy for two guerrilla species: *Elymus repens* (L.) Gould and *Agrostis stolonifera* (L.)? Plant Ecology. 2008;**197**:107–117. DOI: 10.1007/s11258-007-9364-z

[51] Van der Putten WH. Plant defense belowground and spatiotemporal processes in natural vegetation. Ecology. 2003;**84**:2269–2280. DOI: 10.1890/02-0284

[52] Su JS, Jing GH, Jin JW, Wei L, Liu J, Cheng JM. Identifying drivers of root community compositional changes in semiarid grassland on the Loess plateau after long-term grazing exclusion. Ecological Engineering. 2017;**99**:13–21. DOI: 10.1016/j.ecoleng.2016.11.050

[53] Smith NG, Schuster MJ, Dukes JS. Rainfall variability and nitrogen addition synergistically reduce plant diversity in a restored tall grass prairie. Journal of Applied Ecology. 2016;**53**:579–586. DOI: 10.1111/1365-2664.12593

[54] Lamb EG, Stewart AC, Cahill JF Jr. Root system size determines plant performance following short-term soil nutrient pulses. Plant Ecology. 2012;**213**:1803–1812. DOI: 10.1007/s11258-012-0135-0

[55] Wan SQ, Hui DF, Wallace L, Luo YQ. Direct and indirect effects of experimental warming on ecosystem carbon processes in a tall grass prairie. Global Biogeochemical Cycles. 2005;**19**. DOI: 10.1029/2004GB002315

[56] Shahzad T, Chenu C, Repinçay C, Mougin C, Ollier JL, Fontaine S. Plant clipping decelerates the mineralization of recalcitrant soil organic matter under multiple grassland species. Soil Biology and Biochemistry. 2012;**51:**73–80. DOI: 10.1016/j.soilbio.2012.04.014

[57] Zhang W, Parker K, Luo Y, Wan S, Wallace L, Hu S. Soil microbial responses to experimental warming and clipping in a tall grass prairie. Global Change Biology. 2005;**11:**266–277. DOI: 10.1111/j.1365-2486.2005.00902.x

[58] Chen F, Weber KT. Assessing the impact of seasonal precipitation and temperature on vegetation in a grass-dominated rangeland. The Rangeland Journal. 2014;**36**:185–190. DOI: 10.1071/RJ13098

[59] Cleland EE, Chiariello NR, Loarie SR, Mooney HA, Field CB. Diverse responses of phenology to global changes in a grassland ecosystem. Proceedings of the National Academy of Sciences. 2006;**103**:13740–13744. DOI: 10.1073/pnas.0600815103

[60] Bradford J, Lauenroth W, Burke I, Paruelo JM. The influence of climate, soils, weather, and land use on primary production and biomass seasonality in the US Great Plains. Ecosystems. 2006;**9**:934–950. DOI: 10.1007/s10021-004-0164-1

[61] De Boeck HJ, Bassin S, Verlinden M, Zeiter M, Hiltbrunner E. Simulated heat waves affected alpine grassland only in combination with drought. New Phytologist. 2016;**209**:531–541. DOI: 10.1111/nph.13601

[62] Reyes-Fox M, Steltzer H, Trlica MJ, McMaster GS, Andales AA, LeCain DR, et al. Elevated CO_2 further lengthens growing season under warming conditions. Nature. 2014;**510**:259–262. DOI: 10.1038/nature13207

[63] Grime JP, Brown VK, Thompson K, Masters GJ, Hillier SH, Clarke IP, et al. The response of two contrasting limestone grasslands to simulated climate change. Science. 2000;**289**:762–765. DOI: 10.1126/science.289.5480.762

[64] Bokhorst SF, Bjerke JW, Tømmervik H, Callaghan TV, Phoenix GK, et al. Winter warming events damage sub-Arctic vegetation: Consistent evidence from an experimental manipulation and a natural event. Journal of Ecology. 2009;**97**:1408–1415. DOI: 10.1111/j.1365-2745.2009.01554.x

[65] Luedeling E, Guo L, Dai J, Leslie C, Blanke MM. Differential responses of trees to temperature variation during the chilling and forcing phases. Agricultural and Forest Meteorology. 2013;**181**:33–42. DOI: 10.1016/j.agrformet.2013.06.018

[66] Guo L, Dai J, Ranjitkar S, Xu J, Luedeling E. Response of chestnut phenology in China to climate variation and change. Agricultural and Forest Meteorology. 2013;**180**:164–172. DOI: 10.1016/j.agrformet.2013.06.004.

Chapter 5

Leaf Ecology and Radiocesium Contamination in Trees/ Forests

Toshihiro Yoshihara

Additional information is available at the end of the chapter

http://dx.doi.org/10.5772/intechopen.69082

Abstract

Nonessential elements enter/accumulate in trees at certain ratios via the same uptake/translocation systems as essential elements. This phenomenon may not only damage the ecosystem but also result in human health problems. As one such nonessential element, the fate of radiocesium in trees has been extensively studied after the nuclear accident at Fukushima in 2011. Here, to better our understanding of the fate of radiocesium in nature and contribute to plan countermeasures, a review based on recent data for the Fukushima accident will be explicated with historical experiences of the global fallout, the Chernobyl accident, and many laboratory studies. In particular, the effects of specific leaf ecology (deciduous and evergreen), types of radiocesium exposure (dry/wet depositions or root uptake), and decomposition of litter on the fate of radiocesium will be precisely described with a specific uptake/translocation system of potassium, which can be recognized as the most possible entrance of radiocesium into trees.

Keywords: tree, radiocesium, nutrient, potassium, senescence

1. Introduction

1.1. Background of the topic and expected impact

There have been growing concerns about the fate of radionuclides in forests following the accident at the Fukushima Daiichi nuclear power plant (FDNPP) in March 2011. Radiocesium (^{134}Cs and ^{137}Cs, rCs) contamination is of particular concern, because of its comparatively long physical half-lives (more than 2 and 30 years, respectively) and their abundance in the Fukushima fallout (e.g., [1]). Even though cesium (Cs) is a nonessential element, it is true that Cs can enter the plant body via roots and/or leaf surface and get mixed in the natural circulation. In addition, tree ecology, defoliation, would complicate the fate of rCs in forests through the abundance and degradation of litter fall (**Figure 1**). This may affect both the level of internal exposure to persons who take forest products and external exposure to persons who live with forests. To clear the fate of rCs may contribute not only to establish the countermeasure of the present accident, such as decontamination and mitigate actions against radiation damage, but also to develop a novel process to regulate nuclear policies, such as probabilistic risk assessment (PRA).

1.2. Defoliation of senescent leaves

Defoliation of senescent leaves, together with flowering, is the most conspicuous and important phenomenon in tree ecology. Particularly in deciduous species, defoliation is an event that occurs all at once every year in autumn, and thus provides a basis for tree classification. However, defoliation of senescent leaves is not only a seasonal event in tree ecology/physiology but also indicates the positive ability of trees to adapt to limitations of the environment, either climatic or competitive [2]. Deciduous species evolved to shed their leaves to minimize detrimental environmental effects, such as drought or cold stress. By contrast, evergreen species maintain their leaves as long as possible [3]. Interestingly, leaf longevity varies with environmental conditions (e.g., light intensity) even in the same species. Such prolonged leaf longevity helps evergreen species save energy in the development of new leaves [4]. However, evergreen species shed their leaves eventually. Some experience short leaf longevity of less

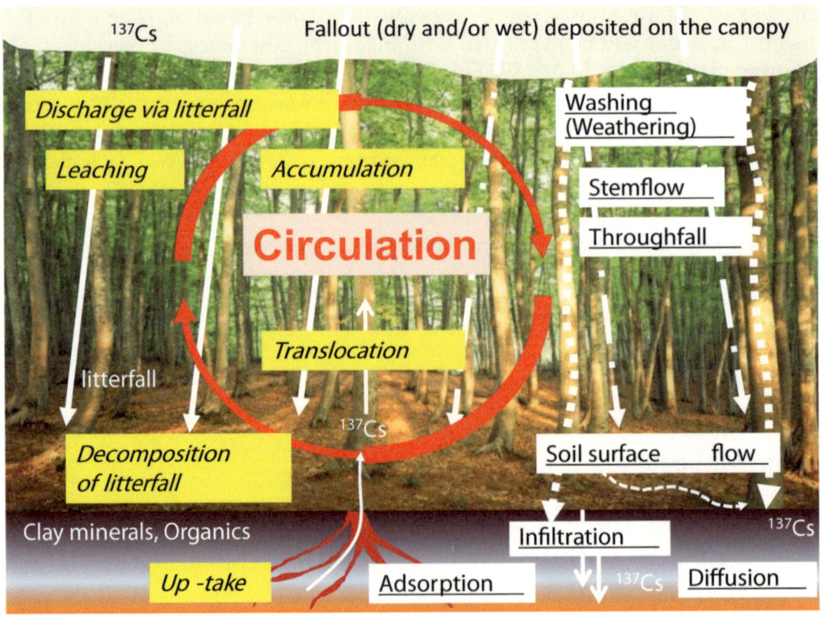

Figure 1. A possible circulation and translocation of radiocesium in forest vegetation is described with related natural forces. The terms written in italics (Accumulation, Leaching, Translocation, Discharge, Decomposition, and Uptake) and those with an underline (others) are showing forces related to biological activities and meteorological/geological activities, respectively.

than 1 year. This is a necessary and constructive step in the life cycle of trees, which is required for the renewal of senescent parts, optimization of the spatial arrangement of leaves, and as a competitive measure against neighboring individuals [4].

1.3. Translocation of essential elements in trees

Before defoliation, trees essentially translocate (i.e., reabsorb) nutrients (i.e., essential elements) from senescent leaves to the tree body or newly developed portions prior to shedding their leaves irrespective of the leaf habit [5, 6]. The translocated nutrient is recycled to develop new leaves and other parts. This is a necessary trait of trees particularly in natural ecosystems, where nutritional resources are poor [7]. It is well known that the efficiency of translocation is dependent on both the type of element and tree species. In this regard, the extended leaf longevities of evergreen species appear to be better than the yearly defoliation of deciduous species. The direct delivery of nutrients from senescent leaves to newly developing parts in evergreen species diminishes the chances of nutritional loss and ensures the growth more efficiently [5, 7–11].

1.4. Translocation of nonessential elements in trees

Nonessential elements also accumulate/translocate in trees at certain ratios, although nonessential elements are often harmful for tree growth. The uptake of these substances is considered to be incidental, occurring via the same uptake/translocation systems as essential elements [12]. For example, cadmium (Cd) is known to be easily translocated from the soil to any part of the trees. Although the possible uptake pathways of Cd are comparatively varied with respect to those of major divalent metal cations, most of these pathways are the same as those of zinc (Zn) and Fe. The reason why these elements share the same uptake pathways has been explained by their physical (e.g., Cd and Fe have a similar ionic radius) and/or chemical (e.g., Cd belongs to the 12th group below Zn) properties. As with the uptake/translocation of essential elements, the uptake/translocation efficiency of nonessential elements is highly dependent on both the types of element and tree species. Relationships between Cd and divalent metal ions, such as Zn and Fe, have been particularly well documented to date from ecological to molecular levels; however, there is still limited knowledge on the relationships of other nonessential elements [12].

1.5. Biological analog of Cs

Because Cs is a nonessential element, it is necessary to take into consideration its relationship with essential elements, when assessing the fate of radiocesium in forests. Potassium (K) is the most important biological analog of Cs, and the metabolism for Cs and K in trees is closely related (**Figure 2**) [13]. For example, sufficient K fertilization can decrease rCs accumulation in trees, whereas K deficiency may increase accumulation [14, 15]. In fact, an increase in K fertilization is one of the most efficient countermeasures for reducing rCs contamination in rice [16, 17]. However, in forest ecosystems, forest soils tend to be K deficient, although the level of deficiency varies widely with seasons and individuals, and it is difficult to apply K fertilization (e.g., [18]). This may emphasize the importance of K recycling, which may affect the status of rCs in trees. On the other hand, correlation between rCs and K status in litter fall is a

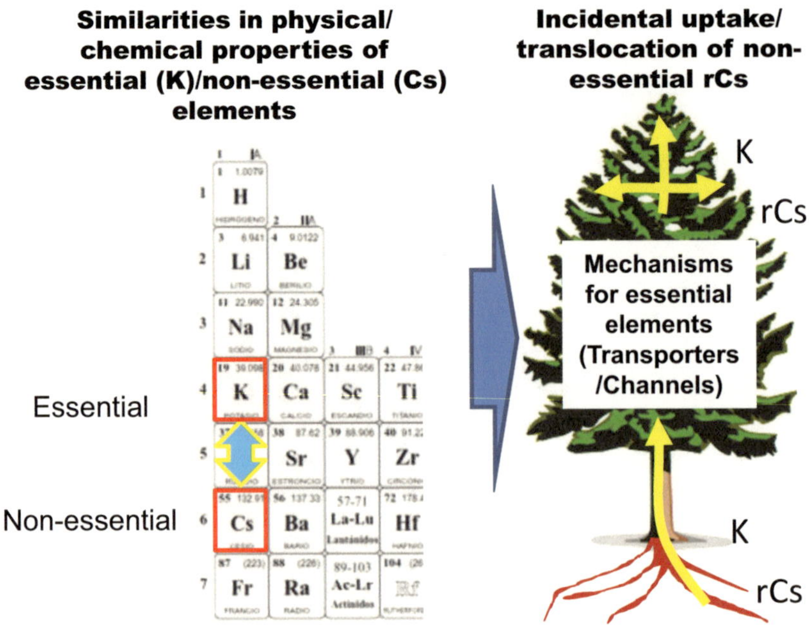

Figure 2. Specific correlation between cesium and potassium governs their accumulation in trees. Potassium (K) is the most important biological analog of Cs, and the metabolism for Cs and K in trees in closely related.

highly species-specific [13, 19–21]. This may reflect the prolonged leaf longevity of individual tree species and the related physiology.

The objective of this study is to gain our understanding of the fate of radiocesium in nature to contribute to plan countermeasures. An explication of recent data for the Fukushima accident with historical experiences of the global fallout, the Chernobyl accident, and many laboratory studies, may help to clarify each universality and/or the particularity. Especially, the effects of three major factors influencing the fate of rCs in forests, types of radiocesium exposure (dry/wet depositions or root uptake), climate, and specific leaf ecology (tree species) on the fate of radiocesium, are precisely described.

2. Materials and methods

We conducted a literature search for data describing changes in rCs in forests/trees with relevant key words, which are necessary to recognize the effects of three major factors as mentioned above. For example, weathering effect, velocity/direction of wind, amount/intensity of precipitation, particle size/chemical form of rCs fallout, sticking effect, leaf surface characteristic, tree height, topology in forests, and stomatal aperture were used to search the effect of the effects of dry/wet depositions. Soil type, humus/humic acid/fulvic acid, clay, mineral, organic matter, polar/boreal/temperate/tropical/desert, and topology in forests were used to search the effect of climates. Translocation/uptake ability of rCs, transfer factor, canopy density, branching geometry, litter fall, decomposition, mycorrhiza, symbiosis, microbial, and transporters/channels were used to search the effect of tree species.

3. Result and discussion

3.1. An overview of the Fukushima accident

3.1.1. The impact of the Fukushima accident on forests/trees

A large earthquake and tsunami struck northeastern Japan on March 11, 2011, and resulted in an accident at the Fukushima Daiichi Nuclear Power Plant (FDNPP).The substantial amount of radionuclides, radioiodine (^{131}I) and radiocesium (^{134}Cs and ^{137}Cs; rCs), contaminated large areas of northern Japan [22–24]. The total amount of ^{137}Cs released from the FDNPP accident to the atmosphere was approximately 12 PBq, 14% (86 PBq) of that from the Chernobyl Nuclear Power Plant (CNPP) accident in April 1986 [25]. The initial radionuclide fallout was observed on March 15, 2011, after the first hydrogen explosion of the reactors. In addition, the peaks of the fallout were observed accordingly with emitted type of radionuclides, observed areas, wind direction/velocity, and precipitation [1]. For example, in Fukushima, the majority of the ^{131}I fallout was observed on March 16, 2011, as dry deposition, whereas that of the ^{137}Cs fallout was observed on March 21, 2011, with rainfall.

Most (approximately 80%) of the radionuclides were directly transported by wind to the Pacific Ocean after discharge into the atmosphere, while the rest was deposited over land [22, 26, 27]. It is worth noting that the portion deposited over land was larger for ^{137}Cs than ^{131}I (22% of ^{137}Cs and 13% of ^{131}I) [1], and the majority was captured by forests [27]. Estimation from earlier surveys indicates that forests occupied at least 1343 km^2 of the total 1778 km^2 contaminated with more than 5 mSv y^{-1} [28]. Another estimation indicates that approximately 70% of the high levels of radioactive fallout (>1000 kBq m^2) was over forested areas, and 21 Tg-DW of forested components were contaminated [29]. The fallout capture ratio in forests seemed to be in proportion to the ratio of the forest occupancy to the total contaminated land areas. In fact, forests extend over approximately 70% of the land area of the Fukushima Prefecture (covering 975,000 ha of 1,378,000 ha) [30, 36]. In addition, forests are considered to be effective interceptors of radionuclides than other land-use areas due to the large surface areas of tree leaves and the height of the trees themselves (e.g., [31, 32]). However, deciduous species had no leaves at the time of the accident, as it occurred during late winter, whereas evergreen species had kept their leaves green. This resulted in a similar initial contamination status in both the Fukushima and Chernobyl cases, in which significant differences among species were noted not only in their canopies but also in the soils under the trees [33, 34]. Kato et al. [35] also estimated a higher level of ^{137}Cs interception by the canopies of Japanese cedar forests (approximately 60% of the total deposition on trees). The evergreen Japanese cedar is the most popular commercial woody species in Japan [30]. In Fukushima Prefecture, Japanese cedar is planted across 65% of plantation forests [36]. It is notable that both the forest occupancy ratio and the specific configuration differ between the FDNPP and the CNPP accidents (forests extend over approximately 40% mainly covered with red pine and spruce in case of the CNPP accidents) in addition to the differences in climate, soil texture, and radionuclides themselves. Thus, much attention should be paid to the fate of radionuclides [37].

3.1.2. The contamination status quo in forests/trees of the disaster areas

After canopy interception, ^{131}I physically decayed within a few months, and major portions of the longer half-lived rCs fallout transferred from the canopy to the soil over time [38–42]. For example, 22–44% of total ^{137}Cs fallout was noted in a forest canopy of Japanese cedar in Fukushima, and 56–78% was in the forest floor around 6 months after the deposition [34]. By contrast, another forest canopy in a similar situation reduced the occupancy ratio to 6% while the forest floor increased that to 74% 4 years later [42]. These observations are not always consistent with models based on the Chernobyl experience, in that a proportion of the radionuclides within the tree parts of forests would disappear within 5 years at most in Fukushima (e.g., [43]). It is possible that differences in climate, topology, and vegetation between Chernobyl and Fukushima affect this discrepancy. In any case, the changes in rCs fallout in the forest canopy can be expressed as an exponential decrease [21, 35, 44, 45]. Exponential decreases in ^{137}Cs concentrations deposited on trees have been commonly observed in the Chernobyl case and under experimental conditions [46–48]. However, the expected half-lives from the equations were highly variable between models even when the models were based on the same data [45]. The model construction may affect this. Thus far, three models have been used to describe the exponential decrease: single exponential, double exponential, and offset exponential. The single exponential model is useful for comparison with values from previous studies, whereas the double exponential model is suitable to show the decrease in loss rates over time. The offset exponential model can demonstrate initial rapid loss with an unattainable residue [45, 49]. In addition, Kato et al. [45] speculated that an estimation caused by a lack of initial data during the first few months from the deposition may have caused the variability. In this regard, the fact that the data collection started 2–6 months after the FDNPP accident and 1–2 years after the CNPP accident is an important point to note when interpreting the model. On the other hand, the decrease in rCs in the canopy is always recognized as occurring in three phases irrespective of the models: "early phase" or "acute reduction phase" (1–3 months), "medium-term phase" (2–3 years), and "long-term phase" or "quasi-stable phase" (3–10+ years) [50]. Sometimes, the phases can be divided into two phases with the combination of "acute reduction phase" and "medium-term phase" or "medium-term phase" and "long-term phase" [51]. Subsequently, rCs fallout in forests is presently in a quasi-stable phase with a kind of elemental circulation including root uptake [42].

Possible factors affecting rCs translocation should be separately considered with respect to the phase of the decrease [44, 50]. In the "acute phase," the translocation of rCs from the canopy to the soil is mainly governed by a rapid mechanical washing with rain and wind. Thus, the factors involved are the amount of precipitations and direction/velocity of winds. These are important not only at the time of deposition but also during the entire "acute phase." This mechanical washing is often called "weathering." In addition, the method of depositions (i.e., dry/wet) is also an important factor in this phase. After this phase, a contribution of biological redistribution processes, such as litter fall, increased in the "medium-term phase" over time. Additionally, the translocation of rCs from tree parts, which receive the direct deposition, to the other parts, which vigorously grow after the deposition, plays a major role in this phase. This translocation process does not actually reduce the amount of rCs from trees, but rather decreases the concentration, and thus is referred to as a "growth dilution" [52]. Root uptake of rCs is still not obvious in this phase; however, it becomes increasingly more evident over time [42, 51, 52]. The major factors in the "medium-term phase" are the tree species and their living conditions, such as soil conditions and climatic conditions of the ecosystem. It is worth noting that the mechanism of uptake/translocation of K is particularly important in this phase (including the molecular aspects). In the "long-term phase," the root uptake of rCs becomes a main player, although other biological redistribution processes still play a major roll. These processes work interactively to establish rCs equilibrium in the forest. Major factors influencing the fate of rCs contamination in forests/trees, particularly regarding the effects of dry/wet depositions, climates, and tree species, are discussed in detail in the following sections.

3.2. Major factors influencing the fate of radiocesium contamination in forests/trees

The results of literature search are summarized in **Table 1** for each of the picked-up three major factors (depositions, climate, and tree species) with the specifications and cofactors. Detailed explanations and discussions are described in the following sections.

Major factors	Specification	Co-factors
➢ Depositions (The determinant of the total amount of contamination)	• Particulate or gaseous matter (dry depositions) • Water-dissolved matter (wet depositions) • Biological traits of trees	• Wind velocity/direction, rCs Particle size • Precipitation amount/Intensity • Leaf area and surface characteristics, Total biomass, Tree heights, Stomatal aperture/respiration, Tree physiology
➢ Climate (The determinant of rCs mobility via generation of specific soils and topologies in forests)	• Polar and boreal, temperate, and tropical	• Meteorological factors (e.g., temperature, moisture, precipitation amount/Intensity, freezing and previous precipitation), Soil texture (Cs-fixing minerals), Vegetation
➢ Tree species (The leading part of rCs circulation in forests)	• Canopy density and branching geometry • Litter fall and decomposition • Interaction with mycorrhiza • Root uptake/Translocation	• Humidity, Light quantity, Forest topography • Leaf longevity, Degradability, Saprophyte activity • Plant species-soil conditions combinations • K pathway (Transporters and channels), Soil conditions

Table 1. Three major factors influencing the fate of radiocesium in forests.

3.2.1. Dry/wet depositions: types of the initial radiocesium fallout

The first contact of radiocesium fallout with forests/trees can be divided into two types based on the physicochemical statuses: depositions as particulate or gaseous matter (dry depositions) and depositions as water-dissolved matter (wet depositions). In other words, the fate of rCs contamination in forests/trees is primarily influenced by these differences through the total amount of and accessibility to the initial deposition. In general, meteorological conditions, such as velocity/direction of wind and amount/intensity of precipitation, are of great importance for dry and wet depositions, respectively [53, 54]. For example, the interception fraction of rCs decreases with increasing amount/intensity of precipitation for wet depositions [49, 55]. The absorption coefficient of dry depositions on wet surfaces drastically increases with an increase of a sticking effect [56]. In addition, the absorption coefficient considerably decreases with increasing particle size of rCs fallout. It is probable that larger particles roll off the plant surface more easily than smaller ones [57].

Biological traits of trees, such as the leaf area (or total biomass), leaf surface characteristics, tree heights, and topology in forests, are also important in the establishment of depositions in both dry/wet forms. Such biological traits directly affect the interception of rCs by trees via changes in micro-weather in forests and/or physiological interactions with rCs [58, 59]. In general, the larger the leaf area, the higher the leaf-generating densities, and the higher the tree height, the greater the amount of rCs interception can be irrespective of the type of the initial fallout [59]. In addition, particularly in gaseous matter deposition, the stomatal aperture also plays a key role [57]. This means that gaseous matter deposition can directly enter trees via stomata by means of respiration. The efficiency of radiocesium interception is affected by the physiological status of trees, such as water demand and/or growth vigor. Interestingly, rCs particles can also enter via the stomata; spinach can intercept more rCs particulate in spring than in summer when the stomata are more open [57]. On the other hand, the water-holding (storage) capacity of leaves is closely related to the interception of rCs through rain [59]. Interception of rain can be quantified in terms of the thickness of a water film covering the foliage (i.e., the surface tension holding the water), and thus a larger capacity may increase interception of rCs [55]. Similarly, the leaf-generating angles and the specific ratio in mass per unit leaf area are also related to the interception of rCs [60].

It is worth noting that these first steps of rCs interception are common among radionuclides and other non-radioactive air-borne pollutants [61]; however, differences in the exact details are evident depending on the elements and chemical forms. The major factor driving such differences is chemical valency. For example, the largest mass interception factor among major radionuclides detected in the same rainfall event after the Chernobyl fallout was observed for ^{140}Ba followed by that for ^{137}Cs, ^{131}I, and ^{106}Ru [62]. Because plant surfaces are negatively charged, the initial retention of anions such as iodide is less than that of polyvalent cations. Similarly, the bivalent cation barium can more easily be retained on plant surfaces than monovalent cesium cation [63]. In this context, Hoffman et al. [64] demonstrated that the mass interception factors for cations (Cd^{2+}, Be^{2+}, Cr^{3+}, Sr^{2+}, Ce^{3+}) are approximately a factor of three to five times higher than for anions ($SO4^{2-}$, I^{-}).

3.2.2. Climate: the contribution to the generation of specific soils and forest topologies

Meteorological conditions relate to all steps of rCs interception. The importance of the meteorological conditions has already been shown for the first steps of the interception (i.e., establishment of depositions in dry/wet forms) in the previous section. The contribution of these conditions is discussed for specific forest ecosystems through dominant vegetation and their related rCs translocation/uptake systems in the following section. The contribution of these conditions and their indigenous periodical changes climate is generally described for three major climate zones (i.e., polar and boreal, temperate, and tropical) in terms of the generation of specific soils and forest topologies influencing the fate of rCs. The influence of moisture condition is also discussed (desert) as a matter of tropical zones.

Soils are mixed products of mineral ores and biological debris [65, 66]. The debris can be chemically separated into three parts: humus, humic acid, and fulvic acid. These mainly originate from components of previously dominant plants. Thus, litter fall and tree bodies, both of which accumulate rCs, also become specific soils and are incorporated into the natural elemental circulation. It is probable that the mixing of organic residues with Cs-fixing minerals is a key process in Cs mobility [15, 67]. In this regard, Rigol et al. [68] demonstrated that rCs adsorption generally does not correlate with the organic matter content in soils, although organic soils with more than 95% organic matter do affect adsorption. On the other hand, the time-dependent pattern of the exchangeable fraction may relate to soil–plant transfer dynamics [68].

The importance of topographical factors in the fate of radiocesium is particularly evident in mountain areas [52]. Two topographical factors are often recognized by the time dependency: "fast hydrological component" or "direct run-off," and "slow erosional component" or "delayed removal." Both factors are mainly based on the natural dispensation that water accumulates in depressions accompanied by a rich amount of rCs deposition [69]. Furthermore, these originally resulted from the influence of rainfall intensity and soil permeability, which depend on soil texture and climatic factors (e.g., freezing and previous precipitation).

(a) **Polar and boreal zones:** Soils in polar area are essentially premature owing to a thick layer of peat (top 10 cm of tundra) due to limited decomposition of plant litter by low temperatures and wet and anaerobic conditions [70]. This thick peat layer blankets the underlying mineral soil [71] and prevents rCs from binding to minerals [72]. In addition, both the diffusion coefficient and the convection velocity of rCs observed in the soils of a polar zone (Antarctic area) are smaller than those determined in temperate zones [73]. On the other hand, the low K_d values of rCs due to the low clay content and high NH^{4+} concentration in the soil solution are responsible for the high soil–plant transfer in organic soils [68]. The combination of these two factors might result in a higher soil–plant transfer of rCs in polar areas than in others.

In boreal zones, forest areas cover 1135 Mha of the total land areas and stock 272 Pg carbon (C) (approximately 32% of the total global C stocks) [74]. In particular, the C stock occupation ratios in biomass are considerably higher than those in soils. This may be due to limited decomposition of plant litter by low temperatures, which enables these zones to provide a major source of humus capable of adsorbing large amounts of rCs [69, 75, 76]. This trait should affect the radionuclide cycle in ecosystems [77]. In fact, the duration of a complete radionuclide cycle in ecosystems in northern regions of European portion of the Russian Federation (the White Sea) is 10 half-life periods, while that in the southern regions (the Black sea) is 2.5 half-life periods [78].

The importance of topographical factors in alpine areas is evident as in other mountain areas [52]. In addition, runoff phenomena in boreal forest ecosystems constitute a dominant factor, particularly just after deposition [79]. The thick organic soil layer, mentioned above, acts as an rCs-impermeable soil layer, and such a layer tends to favor runoff phenomena [69]. For example, rCs activities were higher in damp peats, bogs/flushes, and peaty grasslands than in the adjacent drier and more mineral soils after the Chernobyl accident [69, 75]. Furthermore, polar and boreal climate zone-specific traits, such as snow fall and freezing, play an important role. For example, Gaare [80] found that the activity of Chernobyl-derived rCs in the vegetation of a wind-exposed snow-free area of Norway was approximately three times higher than on adjacent snow-covered areas. In addition, wind-blown snow distributions significantly control vegetation and topography, and thus rCs distribution, in an arctic tundra basin [81].

(b) **Temperate zones:** Although forested areas (767 Mha) and C stock (119 Pg C) in temperate zones are the smallest among the three climate zones [74], it is probable that forested areas in temperate zones have higher potentials both in fixing C and in decomposing litter than in boreal zones due to their moderate temperature and humidity [82]. However, no consensus has been reached regarding the temperature sensitivity of soil C decomposition [83, 84]. On the other hand, the comparatively larger diversity of tree species in temperate zones than in boreal zones can contribute to differences in rCs-adsorbing capacity in soils through degradation sensitivity [85]. In this regard, forests dominated by evergreen species had significantly lower soil organic matters than that dominated by deciduous or mixed species [82]. Nevertheless, the type of radiocesium cycle in the temperate zones, polar and boreal zones (i.e., organic matter adsorbed-type) or tropical zones (i.e., mineral adsorbed-type) may be dependent on the balance of the capacities of C fixation and degradability in each forest.

The effect of topographical factors is also evident with increases in altitude in the temperate zones [86]. For example, the difference in the deposition between the summit of the 800 m mountains and the coastal areas is mainly explained as a factor of rainfall amounts [87]. Many parts of temperate zones, including the Fukushima disaster areas, contain mountainous areas and have comparatively high rainfall, particularly those associated with typhoons [27, 88, 89], and thus should show similar topographical effects. In fact, Koarashi et al. [90] showed that both leaf-litter materials and litter-associated ^{137}Cs accumulated in large amounts at the bottom of the hill slope owing to a topographical effect. In addition, among this litter, newly shed and less-degraded leaf-litter materials occupied 65% of the total ^{137}Cs inventory.

(c) **Tropical zones:** Ecosystems in the tropical zones consist of rainforests, dry deciduous forest, spiny forests, deserts, and others; however, the descriptions related to tropical zones in this study are related to rainforests, with the exception of a separated description for deserts. Although the total C stock in soils and forested areas in the tropical zones is the largest of that in the three climate zones [74], the surface accumulation of soil organic matter is minimal [82, 91]. In addition, the C stock occupation ratios in tropical zones are higher in biomass than in the soil [74]. This may be due to vigorous plant growth and decomposer activities through the higher temperature and humidity in this zone [92]. As a result, the organic content in surface soils is lower and the fixation of rCs higher than that in other zones [91, 93–96]. It is worth noting that even though the organic matter content in tropical zones is comparatively low, the labile fractions of rCs in the forest soil within the tropical zones were larger than those in the grassland soils, particularly those containing volcanic soil [97]. Although higher microbial activity leads to the decomposition of organic materials, it also increases the feedback of ^{137}Cs from organic horizons to fresh litter, and may be a key function in the recycling and persistence of ^{137}Cs in forest soils [60, 97]. On the other hand, Russell et al. [98] indicated a possibility that bacterial sulfate reduction decreases the adsorption of ^{137}Cs in the soil. Thus, in addition to a balance of litter generation and decomposition, the potential of mineral uptake by plants also plays an important role in the fate of ^{137}Cs in tropical forests.

Desert climate also influences the fate of rCs in soils; however, the pattern of this influence appears to be the reverse of that from the other zones. The limited rainfall in deserts usually leads to reduced growth followed by lower organic matter and clay contents in the soils [99]. Consequently, the fate of ^{137}Cs in deserts may be affected by nonbiological traits. Radionuclide-bound particles in a desert soil, collected in Nevada, US, were predominantly transported by infiltration rather than by bulk-mixing processes, such as freeze/thaw, wetting/drying, and bioturbation. The sandy texture and lower clay content of these soils (relative to those in more temperate environments) increase their hydrologic conductivity and hence their infiltrative transport efficiency [100].

3.3. Tree species: the dominance of the radiocesium circulation in forest ecosystems

Because every process related to rCs circulation in forests is based on tree species-specific translocation/uptake ability of rCs, the most important factor affecting the fate of rCs in forests is tree species [52]. The interspecific differences are usually indicated as TF (soil-to-plant transfer factor) values, which can be utilized as an indicator of translocation/uptake ability for direct comparisons among species [91, 101]. It is known that a 10– to 20–fold range in TF can be seen among species [102, 103]. The interspecific differences are particularly obvious after the second stage of natural forest contamination, when considerable portions of the contaminants drop from the canopy to the soils [51]. Here, species-dependent characteristics related to types of initial rCs fallouts are summarized.

3.3.1. Canopy density and branching geometry

In addition to the importance of canopy density in determining the extent of interception, many environmental conditions under the canopy such as humidity, light quantity, and sometimes topographies of the forest can have an effect [52]. For example, the canopy density of major tree species in Europe can be ordered as follows: *Larix, Pinus, Betula, Quercus, Carpinus, Fagus, Abies,* and *Picea* [104]. This order varies with season; deciduous species always drop their leaves during winter. The branching geometry can be classified into two types: monopodial/centrifugal and sympodial/centripetal [105]. Many conifers belong to the former, which is characterized by indefinite growth of the apical bud, whereas deciduous trees generally belong to the latter, which is characterized by a rapid substitution of the main apical bud to a secondary bud.

3.3.2. *Litter fall and decomposition*

When we consider litter fall as a source of rCs to the forest floor, the concentration of rCs is a very important matter. With this in mind, not only the difference in the amount of initial deposition but also the differences in leaf ecology (e.g., leaf longevity) affect the rCs concentration in litter fall. The difference is particularly large between evergreen and deciduous species. For example, the leaf longevity of evergreen Japanese cedar is usually 4–6 years, and a considerable part of the initial deposition remained on each leaf that received direct deposition at least several years after the Fukushima accident [21, 106]. In addition, nonnegligible part of the residual radiocesium in foliar parts was redistributed to newly developing parts each year [21]. A similar situation can be seen in *Pinus* spp. after the Chernobyl accident and other coniferous species [13, 46], although the retention time and the rate vary by tree species. By contrast, in cases of deciduous species, leaves are renewed each year and did not retain directly received deposition in litter fall even after the first year of the accident in both Chernobyl and Fukushima. Interestingly, some of the litter fall of deciduous species, such as that of cherry trees, indicate seasonal variation (i.e., autumnal decrease) in rCs concentration, while no such significant seasonal variation in litter fall of coniferous species is known [19, 20]. This can be explained as specificity in the translocation of rCs with its biological analog, potassium.

When we consider litter fall as a source of organic matters, which captures rCs in the forest soil (or the surface layer) and prolongs the rCs cycle in the forest ecosystem, degradability (i.e., responsibility against microbial decomposition) of litter fall is important. It is well known that degradability is also dependent on tree species, and coniferous needles show somewhat more tolerance to decomposition than deciduous leaves [69]. Interestingly, bacteria and fungi act as the decomposers (saprophytes), and not only degrade litter fall but also accumulate rCs in the soil surface themselves and retard fixation by minerals [69, 107].

3.3.3. *Interaction with mycorrhiza*

Specific fungal activity correlated with a specific tree species in a relationship known as symbiosis is also important in the rCs cycle in forest ecosystems [52]. These fungi are called mycorrhiza or arbuscular mycorrhizal (AM) fungi and play a major role in the accumulation of elements in natural grown trees through their hyphae, extending to wider areas of surface soil layers than tree roots can [108]. Sometimes, mycorrhizas show a great affinity to radiocesium than plant roots and may act as an active concentrator of radiocesium into trees. However, the enhancement effect on rCs uptake is not always significant [109]. In this regard, recent observations demonstrate that the effect is clear but sometimes inconsistent because of combinations of plant species and soil conditions [110]. In loamy sand and loamy soils, the total ^{137}Cs activity accumulated within the AM host sunflower was 2.4- and 3.2-folds higher than in noninoculated plants, respectively. On the other hand, mycorrhizas themselves also act as rCs accumulators in surface soils rather than plants, bacteria, and other saprophytic fungi [52, 111].

3.3.4. Root uptake/translocation

The main factor determining interspecific differences in rCs contamination in trees is the root uptake/translocation ability of rCs in each species. As indicated in the introduction, the K pathway is the most significant [112]. Thus, the K concentration and K/Cs ratios in soils and tree bodies basically exert a great influence on rCs uptake/transport. However, the influence varies by species due to variations in the individual molecular structure of the K pathway and the resultant disparities in the affinity of Cs to the pathway. The molecular structure of the K pathway has been intensely studied in the decades following the Chernobyl accident. These molecular players in this pathway are mainly divided into two types: K transporters and K channels [112, 113]. To date, many K transporters and K channels have been correlated with rCs uptake/transport [113]. For example, HKT1, which belongs to the KT (K^+transporter)/KUP (K^+uptake permease)/HAK (high-affinity K^+transporter) group of K transporters, is denoted as a Na^+/K^+ symporter and was first recognized as having an affinity to Cs in plants [114, 115]. This group comprises several protein families in each species. For example, in the case of *Arabidopsis thaliana*, 13 members of the KT/KUP/HAK protein family are encoded in the genome [116]. KT/KUP/HAK proteins usually poorly discriminate between potassium, rubidium, and cesium, and thus this protein family plays a very important role in rCs uptake [117]. In particular, the functions of HAK1 and HAK5 have been well documented in plants [118–120]. Another transporter in this group, HKT, is also well documented for Cs uptake particularly in monocotyledonous plants. Interestingly, one of these groups of proteins in wheat, HKT1, has independent binding sites for K and Na, and only the site for K can bind Cs [121].

The first K channel was identified as KAT1, which belongs to a group of voltage-gated K channels [122]. It is worth noting that among this group of proteins, KAT1, KAT2, and AKT2/3 are specifically involved in long-distance K transport [123]. Some other proteins, such as Stelar K^+ outward rectifier (SKOR) and guard cell outward-rectifying K^+ (GORK), are also classified in this group. It is reported that Cs blocks K channel activities; however, it is still uncertain whether or not these channels directly mediate Cs [113]. In addition, other types of proteins (voltage-insensitive channels), including cyclic nucleotide-gated channels (CNGCs) and glutamate receptors (GLRs), are known to be possible K channels and are theoretically considered to be major contributors to Cs uptake in roots [124, 125]. Furthermore, zinc-induced facilitator-Like2 protein (ZIFL2) [126] and low-affinity cation transporter1 (LCT1) [127] have been confirmed to mediate K and Cs influx. Most of these known K transporters and K channels in plants have been identified and assessed using an experimental plant species, *Arabidopsis*, or major crop plants like rice and wheat. Few studies exist for tree species other than some early studies using *Populus* spp. (PtKUP1 [128] and PtHAKs [129]). However, Hosoo et al. [130] recently identified a cDNA sequence of a KUP/HAK/KT transporter (*CjKUP1*) in Japanese cedar (*Cryptomeria japonica*) that was revealed to function as a major K transporter in this species. Although the function of CjKUP1 in Cs uptake has not yet been elucidated, further analysis of similar molecular systems for transporters/channels in other tree species might clarify this aspect.

4. Conclusion and future perspectives

Forests, except those in the vicinity of residential areas, have unfortunately been excluded from the governmental decontamination plan [131]. This is understandable given the prioritization of economic and public health issues; however, forestry occupies a nonnegligible proportion of the Fukushima economy [36]. The potential contaminations of forest products, such as timber, mushrooms, and compost, are major concerns. In addition, forests could be a possible source of "the second pollution" [86]. In particular, discharges of rCs from forests to river systems increase greatly following heavy rains, such as those generated by typhoons [27, 88, 89]. Based on the Chernobyl experience, rCs contamination in forests is predicted to continue at least for additional 10 years [51, 52]. However, there are many disparities between the cases of Chernobyl and the Fukushima accident, with vegetation being a major disparity [132, 133]. In the case of Fukushima, the Japanese cedar (*C. japonica*) and Hinoki cypress (*Chamaecyparis obtusa*) are the most frequently observed species [30, 36], whereas in Chernobyl, the Norway spruce (*P. abies*) and Scots pine (*P. sylvestris*) are the most frequently studied species [134]. This disparity is undoubtedly owing to differences in climate and might induce related changes in the fate of rCs, by means such as organic matter contents in the soils and decomposition. On the other hand, although comparatively fewer contribution of root uptake than that from foliar uptake was observed in the Chernobyl accident (e.g., 0.53% per year of the total ^{137}Cs pool in the soil [47]), it is also true that root uptake increased with time after deposition, particularly after the disappearance of the influence of direct deposition [51, 52]. Thus, the separate quantification of root uptake and direct deposition is the next step in the monitoring of the Fukushima accident. The occurrence of root uptake would be a very species-specific and dependent on soil conditions. Further understanding of these matters is not only important for demonstrating the accumulation and cycling of rCs in a forest ecosystem from an ecological perspective but would also be indispensable for assessing the potential impact on human health and establishing countermeasures.

Acknowledgements

This research was supported in part by a grant-in-aid for Scientific Research from the Japan Society for the Promotion of Science (JSPS, 15H04621).

Author details

Toshihiro Yoshihara

Address all correspondence to: yoshiha@criepi.denken.or.jp

Laboratory of Environmental Science, Central Research Institute of Electric Power Industry (CRIEPI), Abiko, Chiba, Japan

References

[1] Morino Y, Ohara T, Nishizawa M. Atmospheric behavior, deposition, and budget of radioactive materials from the Fukushima Daiichi nuclear power plant in March 2011. Geophysical Research Letters. 2011;**38**:L00G11. DOI: 10.1029/2011GL048689

[2] Leopold A. Senescence in plant development. Science. 1961;**134**:1727-1732. DOI: 10.1126/science.134.3492.1727

[3] Kikuzawa K. A cost-benefit analysis of leaf habit and leaf longevity of trees and their geographical pattern. American Naturalist. 1991;**138**:1250-1263. DOI: 10.1086/285281

[4] Kikuzawa K, Lechowicz MJ. Ecology of leaf longevity. Tokyo: Springer Japan KK; 2011. 147 p. DOI: 10.1007/978-4-431-53918-6

[5] Chapin FS III, Johnson D, McKendrick J. Seasonal movement of nutrients in plants of differing growth form in an Alaskan tundra ecosystem: Implications from herbivory. Journal of Ecology. 1980;**68**:189-209. DOI: 10.2307/2259251

[6] Hagen-Thorn A, Varnagiryte I, Nihlgard B, Armolaitis K. Autumn nutrient resorption and losses in four deciduous forest tree species. Forest Ecology & Management. 2006;**228**:33-39. DOI: 10.1016/j.foreco.2006.02.021

[7] Aerts R, Chapin FS III. The mineral nutrition of wild plants revisited: A re-evaluation of processes and patterns. Advances in Ecological Research. 1999;**30**:1-67. DOI: 10.1016/S0065-2504(08)60016-1

[8] Aerts R. The advantages of being evergreen. Trends in Ecology & Evolution. 1995;**10**:402-407. DOI: 10.1016/S0169-5347(00)89156-9

[9] Miyaura T, Hozumi K. A growth model of a single sugi (*Cryptomeria japonica*) tree based on the dry matter budget of its aboveground parts. Tree Physiology. 1993;**13**:263-274. DOI: 10.1093/treephys/13.3.263

[10] Tsubouchi A, Maekawa T, Hiyoshi S, Ueyama Y, Hisajima T. Seasonal concentration changes of various composition in plant leaf (3) Concentrations of various compositions in Japanese cedar's needle. Bulletin of Fukui Prefecture Institute of Environmental Science. 1993;**23**:53-62

[11] Fujimoto S, Kimura S. Divergences in geographical distribution pattern and ecological leaf characteristics between conifers and broad-leaved tall trees: A cost-benefit analysis with special attention to leaf arrangement in a vegetative shoot. Bulletin of the Research Forest, Faculty of Agriculture, Shizuoka University. 2011;**35**:131-152

[12] Clemens S. Toxic metal accumulation, responses to exposure and mechanisms of tolerance in plants. Biochimie. 2006;**88**:1707-1719. DOI: 10.1016/j.biochi.2006.07.003

[13] Ronneau C, Sombre L, Myttenaere C, Andre P, Vanhouche M, Cara J. Radiocesium and potassium behaviour in forest trees. Journal of Environmental Radioactivity. 1991;**14**:259-268. DOI: 10.1016/0265-931X(91)90032-B

[14] Zhu Y, Shaw G, Nisbet A, Wilkins B. Effect of potassium starvation on the uptake of radiocesium by spring wheat (*Triticum aestivum* cv. Tonic). Plant Soil. 2000;**220**:27-34. DOI: 10.1023/A:1004789915175

[15] Kruyts N, Delvaux B. Soil organic horizons as a major source for radiocesium biorecycling in forest ecosystems. Journal of Environmental Radioactivity. 2002;**58**:175-190. DOI: 10.1016/S0265-931X(01)00065-0

[16] Saito T, Ohkoshi S, Fujimura S, Iwabuchi K, Saito M, Nemoto T, Sato M, Sato M, Yoshioka K, Tsukada H. Effect of potassium application on root uptake of radiocesium in rice. In: Proceedings of International System on Environmental Monitoring and Dose Estimation of Residents after Accident of TEPCO's Fukushima Daiichi Nuclear Power Station. Kyoto: Kyoto University Research Reactor Institute Press; 2012. pp. 165-169

[17] Kato N, Kihou N, Ichihashi H, Yabusaki S, Fujiwara H, Kurishima K, Noda T. Potassium fertilizer and other materials as countermeasures to reduce radiocesium levels in rice: Results of urgent experiments in 2011 responding to the Fukushima Daiichi Nuclear Power Plant accident. Soil Science and Plant Nutrition. 2015;**61**:179-190. DOI: 10.1080/00380768.2014.995584

[18] Tripler C, Kaushal S, Likens G, Todd M. Patterns in potassium dynamics in forest ecosystems. Ecology Letters. 2006;**9**:451-466. DOI: 10.1111/j.1461-0248.2006.00891.x

[19] Yoshihara T, Hashida S-N, Abe K, Ajito H. A time dependent behavior of radiocesium from the Fukushima fallout in Litterfall of Japanese flowering cherry tree. Journal of Environmental Radioactivity. 2014;**127**:34-39. DOI: 10.1016/j.jenvrad.2013.09.007

[20] Yoshihara T, Matsumura H, Hashida S-N. Monitoring of seasonal variations in radiocesium from the Fukushima accident in litterfall of four broad-leaf tree species and its correlation with variation in potassium. Journal of Botanical Science. 2016;**5**:6-14

[21] Yoshihara T, Matsumura H, Hashida S-N, Nakaya K. Radiocesium contamination in living and dead foliar parts of Japanese cedar during 2011-2015. Journal of Environmental Radioactivity. 2016;**164**:291-299. DOI: 10.1016/j.jenvrad.2016.08.010

[22] Chino M, Nakayama H, Nagai H, Terada H, Katata G, Yamazawa H. Preliminary estimation of release amounts of ^{131}I and ^{137}Cs accidentally discharged from the Fukushima Daiichi Nuclear Power Plant into the Atmosphere. Journal of Nuclear Science and Technology. 2011;**48**:1129-1134. DOI: 10.1080/18811248.2011.9711799

[23] Kinoshita N, Sueki K, Sasa K, Kitagawa J, Ikarashi S, Nishimura T, Wong Y, Satou Y, Handa K, Takahashi T, Sato M, Yamagata T. Assessment of individual radionuclide distributions from the Fukushima nuclear accident covering central-east Japan. Proceedings of the National Academy of Sciences of the United States of America. 2011;**108**:19526-19529. DOI: 10.1073/pnas.1111724108

[24] NRA (Nuclear Regulation Authority, Japan). Monitoring Information of Environmental Radioactivity Level. [Internet]. 2012. Available from: http://radioactivity.nsr.go.jp/en/ [Accessed: December 28, 2016]

[25] UNSCEAR (United Nations Scientific Committee of the Effects of Atomic Radiation). Exposures and Effects of the Chernobyl Accident. [Internet]. 2000. Available from: http://www.unscear.org/docs/reports/annexj.pdf [Accessed: March 6, 2017]

[26] Kawamura H, Kobayashi T, Furuno A, In T, Ishikawa Y, Nakayama T, Shima S, Awaji T. Preliminary numerical experiments on oceanic dispersion of ^{131}I and ^{137}Cs discharged into the ocean because of the Fukushima Daiichi Nuclear Power Plant disaster. Journal of Nuclear Science and Technology. 2011;**48**:1349-1356. DOI: 10.1080/18811248.2011.9711826

[27] Evrard, O. Lacebya JP, Lepagea H, Onda Y, Cerdanc O, Ayraulta S. Radiocesium transfer from hillslopes to the Pacific Ocean after the Fukushima Nuclear Power Plant accident: A review. Journal of Environmental Radioactivity. 2015;**148**:92-110. DOI: 10.1016/j.jenvrad.2015.06.018

[28] MOE (Ministry of Environment). The 2nd Skull Session for Environmental Recovery from the Great East Japan Earthquake. [Internet]. 2012. Available from: www.env.go.jp/jishin/rmp/conf/02-mat4.pdf [Accessed: December 28, 2016]

[29] Hashimoto S, Ugawa S, Nanko K, Shichi K. The total amounts of radioactivity contaminated materials in forests in Fukushima, Japan. Scientific Reports. 2012;**2**:416. DOI: 10.1038/srep00416

[30] Forestry Agency. Forests/Timbers and Radionuclides-Revival of Forests and Forestry in Fukushima. [Internet]. 2014. In Japanese. Available from: https://www.pref.fukushima.lg.jp/uploaded/attachment/100065.pdf [Accessed: January 4, 2017]

[31] Bunzl K, Kracke W. Cumulative deposition of ^{137}Cs, $^{238, 239, 240}Pu$, and ^{241}Am from global fallout in soils from forests, grassland and arable land in Bavaria (FRG). Journal of Environmental Radioactivity. 1988;**8**:1-14. DOI: 10.1016/0265-931X(88)90010-0

[32] Petroff A, Mailliat A, Amielh M, Anselmet F. Aerosol dry deposition on vegetative canopies. Part I: Review of present knowledge. Atomic Environment. 2008;**42**:3625-3653. DOI: 10.1016/j.atmosenv.2007.09.043

[33] Yoshihara T, Matsumura H, Hashida S-N, Nagaoka T. Radiocesium contaminations of 20 wood species and the corresponding gamma-ray dose rates around the canopies at 5 months after the Fukushima nuclear power plant accident. Journal of Environmental Radioactivity. 2013;**115**:60-68. DOI: 10.1016/j.jenvrad.2012.07.002

[34] Komatsu M, Kaneko S, Ohashi S, Kuroda K, Sano T, Ikeda S, Saito S, Kiyono Y, Tonosaki M, Miura S, Akama A. Characteristics of initial deposition and behavior of radiocesium in forest ecosystems of different locations and species affected by the Fukushima Daiichi Nuclear Power Plant accident. Journal of Environmental Radioactivity. 2016;**161**:2-10. DOI: 10.1016/j.jenvrad.2015.09.016

[35] Kato H, Onda T, Gomi T. Interception of the Fukushima reactor accident–derived ^{137}Cs, ^{134}Cs and ^{131}I by coniferous canopies. Geophysical Research Letters. 2012;**39**:L20403. DOI: 10.1029/2012GL052928

[36] Fukushima Prefecture. Forests in Fukushima Prefecture. [Internet]. 2016. Available from: http://www.pref.fukushima.lg.jp/sec/36055a/shinrinkeikaku.html [Accessed: December 28, 2016]

[37] AESJ (Atomic Energy Society Japan). Reports of Emergency Survey and Restoration Activity at FDNPP Accident. [Internet]. 2013. Available from: http://www.aesj.or.jp/CU/ [Accessed: January 5, 2017]

[38] Kato H, Onda Y, Teramage M. Depth distribution of ^{137}Cs, ^{134}Cs, and ^{131}I in soil profile after Fukushima Dai-ichi Nuclear Power Plant accident. Journal of Environmental Radioactivity. 2012;**111**:59-64. DOI: 10.1016/j.jenvrad.2011.10.003

[39] Kajimoto T, Saito S, Kawasaki T, Kabeya D, Yazaki K, Tanaka H, Ota T, MatsumotoY, Tabuchi R, Kiyono Y, Takano T, Kuroda K, Fujiwara T, Suzuki Y, Komatsu M, Ohashi S, Kaneko S, Akama A, Takahashi M. Dynamics of radiocesium in forest ecosystems affected by the Fukushima Daiichi nuclear power plant accident: Species-related transfer processes of radiocesium from tree crowns to ground floor during the first two years. Journal of the Japanese Forest Society. 2015;**97**:33-43

[40] Nishikiori T, Watanabea M, Koshikawa MK, Takamatsu T, Ishii Y, Ito S, Takenaka A, Watanabe K, Hayashi S. Uptake and translocation of radiocesium in cedar leaves following the Fukushima nuclear accident. Science of the Total Environment. 2015;**502**:611-616. DOI: 10.1016/j.scitotenv.2014.09.063

[41] Endo I, Ohte N, Iseda K, Tanoi K, Hirose A, Kobayashi NI, Murakami M, Tokuchid N, Ohashi M. Estimation of radioactive 137-cesium transportation by litterfall, stemflow and throughfall in the forests of Fukushima. Journal of Environmental Radioactivity. 2015;**149**:176-785. DOI: 10.1016/j.jenvrad.2015.07.027

[42] Yoschenko V, Takase T, Konoplev A, Nanba K, Onda Y, Kivva S, Zheleznyak M, Sato N, Keitoku K. Radiocesium distribution and fluxes in the typical Cryptomeria japonica forest at the late stage after the accident at Fukushima Dai-Ichi Nuclear Power Plant. Journal of Environmental Radioactivity. 2017;**166**:45-55

[43] Hashimoto S, Matsuura T, Nanko K, Linkov I, Shaw G, Kaneko S. Predicted spatio-temporal dynamics of radiocesium deposited onto forests following the Fukushima nuclear accident. Scientific Reports. 2013;**3**:2564. DOI: 10.1038/srep02564

[44] Loffredo N, Onda Y, Kawamori A, Kato H. Modeling of leachable ^{137}Cs in throughfall and stemflow for Japanese forest canopies after Fukushima Daiichi Nuclear Power Plant accident. Science of the Total Environment. 2014;**493**:701-707. DOI: 10.1016/j.scitotenv.2014.06.059

[45] Kato H, Onda Y, Hisadome K, Loffredo N, Kawamori A. Temporal changes in radiocesium deposition in various forest stands following the Fukushima Dai-ichi Nuclear Power Plant accident. Journal of Environmental Radioactivity. 2017;**166**:449-457. DOI: 10.1016/j.jenvrad.2015.04.016

[46] Bunzl K, Schimmack W, Kreutzer K, Schierl R. Interception and retention of Chernobyl-derived ^{134}Cs, ^{137} Cs and ^{106} Ru in a spruce stand. Science of the Total Environment. 1989;**78**:77-87. DOI: 10.1016/0048-9697(89)90023-5

[47] Goor F, Thiry Y. Processes, dynamics and modelling of radiocaesium cycling in a chronosequence of Chernobyl–contaminated Scots pine (*Pinus sylvestris* L.) plantations. Science of the Total Environment. 2004;**325**:163-180. DOI: 10.1016/j.scitotenv.2003.10.037

[48] Thiry Y, Garcia-Sanchez L, Hurtevent P. Experimental quantification of radiocesium recycling in a coniferous tree after aerial contamination: Field loss dynamics, translocation and final partitioning. Journal of Environmental Radioactivity. 2016;**161**:42-50. DOI: 10.1016/j.jenvrad.2015.12.017

[49] Kinnersley R, Shaw G, Bell J, Minski M, Goddard A. Loss of particulate contaminants from plants canopies under wet and dry conditions. Environmental Pollution. 1996;**91**:227-235. DOI: 10.1016/0269-7491(95)00047-X

[50] Shaw G. Radionuclides in forest ecosystems. Radioactivity Environment. 2007;**10**:127-155. DOI: 10.1016/S1569-4860(06)10006-6

[51] Tikhomirov F, Shcheglov A. Main investigation results on the forest radioecology in the Kyshtym and Chernobyl accident zones. Science of the Total Environment. 1994;**157**:45-57. DOI: 10.1016/0048-9697(94)90564-9

[52] Nimis P. Radiocesium in plants of forest ecosystems. Studia Geobotanica. 1996;**15**:3-49

[53] Anspaugh LR, Steven LS, Konstantin IG, Ilya AL, Reed MM, Sergei MS. Movement of radionuclides in terrestrial ecosystems by physical processes. Health Physics. 2002;**82**:669-679

[54] Madoz-Escande C, Garcia-Sanchez L, Bonhomme T, Morello M. Influence of rainfall characteristics on elimination of aerosols of cesium, strontium, barium and tellurium deposited on grassland. Journal of Environmental Radioactivity. 2005;**84**:1-20. DOI: 10.1016/j.jenvrad.2005.03.006

[55] Kinnersley R, Goddard A, Minski M, Shaw G. Interception of cesium contaminated rain by vegetation. Atmospheric Environment. 1997;**31**:1137-1145. DOI: 10.1016/S1352-2310(96)00312-3

[56] Pinder IJ, Adriano DC, Ciravolo TG, Doswell AC, Yehling DM. The interception and retention of Pu-238 deposition by orange trees. Health Physics. 1987;**52**:707-715

[57] Tschiersch J, Shinonaga T, Heuberger H. Dry deposition of gaseous radioiodine and particulate radiocaesium onto leafy vegetables. Science of the Total Environment. 2009;**407**:5685-5693.

[58] Ould-Dada Z. Dry deposition profile of small particles within a model spruce canopy. Science of the Total Environment. 2002;**282**:83-96. DOI: 10.1016/S0048-9697(01)00965-2

[59] Pröhl G. Interception of dry and wet deposited radionuclides by vegetation. Journal of Environmental Radioactivity. 2009;**100**:675-682. DOI: 10.1016/j.jenvrad.2008.10.006

[60] Vandecasteele C, Baker S, Förstel H, Muzinsky M, Millan R, Madoz-Escande C, Tormos J, Sauras T, Schulte E, Colle C. Interception, retention, translocation under greenhouse conditions of radiocaesium and radiostrontium from a simulated accidental source. Science of the Total Environment. 2001;**278**:199-214. DOI: 10.1016/S0048-9697(01)00649-0

[61] Unsworth M, Wilshaw J. Wet, occult and dry deposition of pollutants on forests. Agricultural and Forest Meteorology. 1989;**47**:221-238

[62] Jacob P, Müller H, Pröhl G, Voigt G, Berg D, Paretzke HG, Regulla D. Environmental behaviour of radionuclides deposited after the reactor accident of Chernobyl and related exposures. Radiation and Environmental Biophysics. 1993;**32**:193-207. DOI: 10.1007/BF01209770

[63] Ertel J, Paretzke H, Ziegler H. ^{137}Cs penetration by contact exchange through isolated plant cuticles: cuticles as asymmetric transport membranes. Plant, Cell & Environment. 1992;**15**:211-219. DOI: 10.1111/j.1365-3040.1992.tb01475.x

[64] Hoffman F, Thiessen K, Rael R. Comparison of interception and initial retention of wet-deposited contaminants on leaves of different vegetation types. Atmospheric Environment. 1995;**29**:1771-1775. DOI: 10.1016/1352-2310(95)00099-K

[65] Yonebayashi K, Hattori T. Chemical and biological studies on environmental humic acids I. Composition of elemental and functional groups of humic Acids. Soil Science and Plant Nutrition. 1988;**34**:571-584. DOI: 10.1080/00380768.1988.10416472

[66] Stevenson FJ. Humus Chemistry. 2nd ed. Chichester: Wiley; 1994. p. 512

[67] Kruyts N, Titeux H, Delvaux B. Mobility of radiocesium in three distinct forest floors. Science of the Total Environment. 2004;**319**:241-252. DOI: 10.1016/S0048-9697(03)00369-3

[68] Rigol A, Vidal M, Rauret G. An overview of the effect of organic matter on soil-radiocaesium interaction: Implications in root uptake. Journal of Environmental Radioactivity. 2002;**58**:191-216. DOI: 10.1016/S0265-931X(01)00066-2

[69] Desmet G, Nassimbeni P, Belli M. editors. Transfer of Radionuclides in Natural and Semi-Natural Environments. London: Elsevier; 1990. p. 707

[70] Jonasson S, Chapin FS, Shaver G. Biogeochemistry in the arctic: Patterns, processes, and controls. In: Schulze ED, Harrison SP, Heimann M, Holland EA, Lloyd JJ, Prentice IC, Schimel D, editors. Global Biogeochemical Cycles in the Climate System. New York, NY: Academic Press. 2001. pp. 139-150

[71] Shaver, G. Carbon turnover in Alaskan tundra soils: Effects of organic matter quality, temperature, moisture and fertilizer. Journal of Ecology. 2006;**94**:740-753. DOI: 10.1111/j.1365-2745.2006.01139.x

[72] Baskaran M, Kelley J, Naidu A, Hoiieman D. Environmental radiocesium in subarctic and arctic Alaska following Chernobyl. Arctic. 1991;**44**:346-350

[73] Schuller P, Bunzl K, Voigt G, Handl J, Ellies A, Castillo A. Fallout radiocesium in an Antarctic region: Deposition history, activity densities and vertical transport in soils. Radiation and Environmental Biophysics. 2002;**41**:295-302. DOI: 10.1007/s00411-002-0169-0

[74] Pan Y, Birdsey RA, Fang J, Houghton R, Kauppi PE, Kurz WA, Phillips OL, Shvidenko A, Lewis SL, Canadell JG, Ciais P, Jackson RB, Pacala SW, McGuire AD, Piao S, Rautiainen A, Sitch S, Hayes D. A large and persistent carbon sink in the world's forests. Science. 2011;**333**:988-993. DOI: 10.1126/science.1201609

[75] Bergman R. The distribution of radioactive caesium in boreal forest ecosystems. Studies in Environmental Science. 1994;**62**:335-379. DOI: 10.1016/S0166-1116(08)71718-7

[76] Guillitte O, Melin J, WaUberg L. Biological pathways of radionuclides originating from the Chernobyl fallout in a boreal forest ecosystem. Science of the Total Environment. 1994;**157**:207-215. DOI: 10.1016/0048-9697(94)90581-9

[77] Gavrikov V, Sharafutdinov R, Mitev A. Turnover of ^{137}Cs in 'soil tree' system: An experience of measuring the isotope flows in a Siberian conifer forest. Journal of Environmental Radioactivity. 2016;**165**:115-123. DOI: 10.1016/j.jenvrad.2016.09.012

[78] Pokarzhevskii A, Uspenskaya E, Filimonova Z. Global radioactive contamination background in terrestrial ecosystems 13 years after the Chernobyl accident. Russian Journal of Ecology. 2003;**34**:73-79. DOI: 10.1023/A:1023077811695

[79] Nylén T, Grip H. Transport of caesium-137 in a forest catchment. In: Feldt W, editor. The radioecology of natural and artificial radionuclides. Proceedings of the XVth Regional Congress. IRPA, Visby, Gotland Sweden Nord; 1989. pp. 221-226

[80] Gaare E. The Chernobyl accident: Can lichens be used to characterize a radiocesium contaminated range? Rangifer. 1987;**7**:46-50. DOI: 10.7557/2.7.2.716

[81] Essery R, Pomeroy J. Vegetation and topographic control of wind-blown snow distributions in distributed and aggregated simulations for an Arctic tundra basin. Journal of Hydrometeorology. 2004;**5**:735-744. DOI: 10.1175/1525-7541(2004)005<0735:VATCOW>2.0.CO;2

[82] Vogt K, Vogt DJ, Brown S, Tilley JP, Edmonds RL, Silver WL, Siccama TG. Dynamics of forest floor and soil organic matter accumulation in boreal, temperate, and tropical forests. Advances in Soil Science. 1995;**62**:159-178

[83] Anderson J. The effects of climate change on decomposition processes in grassland and coniferous forests. Ecological Applications. 1991;**1**:326-347. DOI: 10.2307/1941761

[84] Davidson E, Janssens I. Temperature sensitivity of soil carbon decomposition and feedbacks to climate change. Nature. 2006;**440**:165-173. DOI: 10.1038/nature04514

[85] Cornelissen J. An experimental comparison of leaf decomposition rates in a wide range of temperate plant species and types. Journal of Ecology. 1996;**84**:573-582. DOI: 10.2307/2261479

[86] Likuku A. The influence of topography and vegetation, canopy on the deposition of atmospheric particulates studied with ^{210}Pb and ^{137}Cs soil inventory measurements [Ph.D. thesis]. School of Physics, University of Edinburgh; 2003

[87] Dore A, Choularton T, Fowler D, Storton-West R. Field measurements of wet deposition in an extended region of complex topography. Quarterly Journal of the Royal Meteorological Society. 1990;**116**:1193-1212. DOI: 10.1002/qj.49711649510

[88] Nagao S, Kanamori M, Ochiai S, Tomihara S, Fukushi K, Yamamoto M. Export of ^{134}Cs and ^{137}Cs in the Fukushima river systems at heavy rains by Typhoon Roke in September 2011. Biogeosciences. 2013;**10**:6215-6223. DOI: 10.5194/bg-10-6215-2013

[89] Ueda S. Fluvial discharges of radiocaesium from watersheds contaminated by the Fukushima Dai-ichi Nuclear Power Plant accident. Japanese Journal of Environmental Radioactivity. 2013;**118**:96-104. DOI: 10.1016/j.jenvrad.2012.11.009

[90] Koarashi J, Atarashi-Andoh M, Takeuchi E, Nishimura S. Topographic heterogeneity effect on the accumulation of Fukushima-derived radiocesium on forest floor. Scientific Reports. 2014;**4**:6853. DOI: 10.1038/srep06853

[91] IAEA-TECDOC-1616. Quantification of Radionuclide Transfer in Terrestrial and Freshwater Environments for Radiological Assessments [Internet]. 2009. Available from: http://www-pub.iaea.org/MTCD/Publications/PDF/te_1616_web.pdf [Accessed: March 6, 2017]

[92] Bonan G. Forests and climate change; forcings, feedbacks, and the climate benefits of forests. Science. 2008;**320**:1444-1449. DOI: 10.1126/science.1155121

[93] Rahman M, Voigt G. Radiocaesium soil-to-plant transfer in tropical environments. Journal of Environmental Radioactivity. 2004;**71**:127-138. DOI: 10.1016/S0265-931X(03) 00163-2

[94] Velasco H, Ayub J, Sansone U. Influence of crop types and soil properties on radionuclide soil-to-plant transfer factors in tropical and subtropical environments. Journal of Environmental Radioactivity. 2009;**100**:733-738. DOI: 10.1016/j.jenvrad.2008.12.014

[95] Chiang P. Cesium and strontium sorption by selected tropical and subtropical soils around nuclear facilities. Journal of Environmental Radioactivity. 2010;**101**:472-481. DOI: 10.1016/j.jenvrad.2008.10.013

[96] Twining J, editor. Tropical Radioecology. Oxford, UK: Elsevier; 2012. p. 365

[97] Chiu CY, Wang CJ, Huang CC. Availability and immobilization of ^{137}Cs in subtropical high mountain forest and grassland soils. Journal of Environmental Radioactivity. 2008;**99**:882-889. DOI: 10.1016/j.jenvrad.2007.11.015

[98] Russell R, Holden P, Payne T, McOrist G. The effect of sulfate-reducing bacteria on adsorption of ^{137}Cs by soils from arid and tropical regions. Journal of Environmental Radioactivity. 2004;**74**:151-158

[99] Day A, Ludeke K. Plant Nutrients in Desert Environments. New York, NY: Springer-Verlag; 1993. p. 117

[100] Pelletier J. Geomorphic control of radionuclide diffusion in desert soils. Geophysical Research Letters. 2005;**32**:L23401, DOI: 10.1029/2005GL024347

[101] Ehlken S, Kirchner G. Environmental processes affecting plant root uptake of radioactive trace elements and variability of transfer factor data: A review. Journal of Environmental Radioactivity. 2002;**58**:97-112. DOI: 10.1016/S0265-931X(01)00060-1

[102] Ertel J, Ziegler H. Cs-134/137 contamination and root uptake of different forest trees before and after the Chernobyl accident. Radiation and Environmental Biophysics. 1991;**30**:147-157. DOI: 10.1007/BF01219349

[103] Broadley M, Willey N. Differences in root uptake of radiocaesium by 30 plant taxa. Environmental Pollution. 1997;**97**:11-15. DOI: 10.1016/S0269-7491(97)00090-0

[104] Fraiture A. Introduction to the radioecology of forest ecosystems and survey of radioactive contamination in food products from forests. Commission of the European Communities, Radiation protection. 1992;**57**:1-103

[105] Barthélémy D, Caraglio Y. Plant architecture: A dynamic, multilevel and comprehensive approach to plant form, structure and ontogeny. Annals of Botany. 2007;**99**:375-407. DOI: 10.1093/aob/mcl260

[106] Yoshihara T, Matsumura H, Tsuzaki M, Wakamatsu T, Kobayashi T, Hashida S-N, Nagaoka T, Goto F. Changes in radiocesium contamination from Fukushima in foliar parts of 10 common tree species in Japan between 2011 and 2013. Journal of Environmental Radioactivity. 2014;**138**:220-226. DOI: 10.1016/j.jenvrad.2014.09.002

[107] Vinichuk M, Johanson K, Rosen K, Nilsson I. Role of the fungal mycelium in the retention of radiocaesium in forest soils. Journal of Environmental Radioactivity. 2005;**78**:77-92. DOI: 10.1016/j.jenvrad.2004.02.008

[108] de Boulois H, Joner EJ, Leyval C, Jakobsen I, Chen BD, Roos P, Thiry Y, Rufyikiri G, Delvaux B, Declerck S. Role and influence of mycorrhizal fungi on radiocesium accumulation by plants. Journal of Environmental Radioactivity. 2008;**99**:785-800. DOI: 10.1016/j.jenvrad.2007.10.008

[109] Joner E, Roos P, Jansa J, Frossard E, Leyval C, Jakobsen I. No significant contribution of arbuscular mycorrhizal fungi to transfer of radiocesium from soil to plants. Applied Environmental Microbiology. 2004;**70**:6512-6517. DOI: 10.1128/AEM.70.11.6512-6517.2004

[110] Vinichuk M, Mårtensson A, Ericsson T, Rosén K. Effect of arbuscular mycorrhizal (AM) fungi on ^{137}Cs uptake by plants grown on different soils. Journal of Environmental Radioactivity. 2013;**115**:151-156. DOI: 10.1016/j.jenvrad.2012.08.004

[111] Bakken L, Olsen R. Accumulation of radiocaesium in fungi. Canadian Journal of Microbiology. 1990;**36**:704-710. DOI: 10.1139/m90-119

[112] Zhu YG, Smolders E. Plant uptake of radiocaesium: A review of mechanisms, regulation and application. Journal of Experimental Botany. 2000;**351**:1635-1645. DOI: 10.1093/jexbot/51.351.1635

[113] Shin R, Adams E. Cesium uptake in plants: Mechanism, regulation and application for phytoremediation. In: Gupta DK, Walther C, editors. Impact of Cesium on Plants

and the Environment. Basel: Springer International Publication Switzerland; 2017. pp. 101-124

[114] Schachtman DP, Schroeder J. Structure and transport mechanism of a high-affinity potassium uptake transporter from higher plants. Nature. 1994;**370**:655-658. DOI: 10.1038/370655a0

[115] Gassmann W, Rubio F, Schroeder J. Alkali cation selectivity of the wheat root high-affinity potassium transporter HKT1. Plant Journal. 1996;**10**:869-882. DOI: 10.1046/j.1365-313X.1996.10050869.x

[116] Ahn S, Shin R, Schachtman D. Expression of KT/KUP genes in Arabidopsis and the role of root hairs in K + uptake. Plant Physiology. 2004;**134**:1135-1145.DOI: 10.1104/pp.103.034660

[117] Very A, Sentenac H. Molecular mechanisms and regulation of K+ transport in higher plants. Annual Review of Plant Biology. 2003;**54**:575-603. DOI: 10.1146/annurev.arplant.54.031902.134831

[118] Rubio F, Santa-María G, Rodríguez-Navarro A. Cloning of Arabidopsis and barley cDNAs encoding HAK potassium transporters in root and shoot cells. Physiology Planta. 2000;**109**:34-43. DOI: 10.1034/j.1399-3054.2000.100106.x

[119] Qi Z, Hampton CR, Shin R, Barkla BJ, White PJ, Schachtman DP. The high affinity K+ transporter AtHAK5 plays a physiological role in planta at very low K+ concentrations and provides a caesium uptake pathway in Arabidopsis. Journal of Experimental Botany. 2008;**59**:595-607. DOI: 10.1093/jxb/erm330

[120] Aleman F, Caballero F, Ródenas R, Rivero RM, Martínez V, Rubio F. The F130S point mutation in the Arabidopsis high-affinity K+ transporter AtHAK5 increases K+ over Na+ and Cs+ selectivity and confers Na+ and Cs+ tolerance to yeast under heterologous expression. Frontiers in Plant Science. 2014;**5**:430. DOI: 10.3389/fpls.2014.00430

[121] Schachtman D. Molecular insights into the structure and function of plant K + transport mechanisms. Biochimica et Biophysica Acta. 2000;**1465**:127-139

[122] Sentenac H, Bonneaud N. Cloning and expression in yeast of a plant potassium ion transport system. Science. 1992;**256**:663. DOI: 10.1126/science.1585180

[123] Adams E, Shin R. Transport, signaling, and homeostasis of potassium and sodium in plants. Journal of Integrative Plant Biology. 2014;**56**:231-249. DOI: 10.1111/jipb.12159

[124] Hampton C, Bowen HC, Broadley MR, Hammond JP, Mead A, Payne KA, Pritchard J, White PJ. Cesium toxicity in Arabidopsis. Plant Physiology. 2004;**136**:3824-3837. DOI: 10.1104/pp.104.046672

[125] Kanter U, Hauser A, Michalke B, Dräxl S, Schäffner AR. Caesium and strontium accumulation in shoots of Arabidopsis thaliana: Genetic and physiological aspects. Journal of Experimental Botany. 2010;**61**:3995-4009. DOI: 10.1093/jxb/erq213

[126] Remy E, Cabrito TR, Batista RA, Teixeira MC, Sá-Correia I, Duque P. The major facilitator superfamily transporter ZIFL2 modulates cesium and potassium homeostasis in Arabidopsis. Plant and Cell Physiology. 2015;**56**:148-162. DOI: 10.1093/pcp/pcu157

[127] White PJ, Broadley M. Tansley Review No. 113, Mechanisms of caesium uptake by plants. New Phytology. 2000;**147**:241-256. DOI: 10.1046/j.1469-8137.2000.00704.x

[128] Langer K, Ache P, Geiger G, Stinzing A, Arend M, Wind C, Regan S, Fromm J, Hedrich R. Poplar potassium transporters capable of controlling K+ homeostasis and K+-dependent xylogenesis. Plant Journal. 2002;**32**:997-1009. DOI: 10.1046/j.1365-313X.2002.01487.x

[129] He C, Cui K, Duan A, Zeng Y, Zhang J. Genome-wide and molecular evolution analysis of the Poplar KT/HAK/KUP potassium transporter gene family. Ecological Evolution. 2013;**2**:1996-2004. DOI: 10.1002/ece3.299

[130] Hosoo Y, Kimura Y, Nanatani K, Uozumi N. Molecular cloning and expression analysis of a gene encoding KUP/HAK/KT-type potassium uptake transporter from *Cryptomeria japonica*. Trees. 2014;**28**:1527-1537. DOI: 10.1007/s00468-014-1059-1

[131] MOE (Ministry of Environment). Decontamination Plans for Forests. [Internet]. 2012. Available from: http://josen.env.go.jp/about/efforts/forest.html [Accessed: December 28, 2016]

[132] Yamada M. A brief review of environmental impacts and health effects from the accidents at the three mile islands, Chernobyl and Fukushima Daiichi Nuclear Power plants. Radiation Emergency Medicine. 2012;**1**:33-39

[133] Steinhauser G, Brandl A, Johnson T. Comparison of the Chernobyl and Fukushima nuclear accidents: A review of the environmental impacts. Science of the Total Environment. 2014;**470**:800-817. DOI: 10.1016/j.scitotenv.2013.10.029

[134] Brown C, Ball J. World View of Plantation Grown Wood. [Internet]. 2013. Available from: http://www.fao.org/forestry/4596-0bc06e1b1190f66bb48651bfd756f37fd.pdf [Accessed: January 8, 2017]

Section 3

Functional Ecology

Plant-Microbe Ecology: Interactions of Plants and Symbiotic Microbial Communities

Ying-Ning Ho, Dony Chacko Mathew and
Chieh-Chen Huang

Additional information is available at the end of the chapter

http://dx.doi.org/10.5772/intechopen.69088

Abstract

Plant community dynamics are driven by the microbial mediation of soil resource partitioning and sharing by the inhibition of other host symbionts or sharing the broadly specific symbiotic fungi. The plant phenotype and ecology can be affected by the impact of the symbiotic microbes on the environment and competition for soil resources.

The advent of modern biotechnology has made it easy to study plant-microbe interactions further. Current genomic technologies applied to natural and artificial systems have shown that the plant genotype has a small but significant effect on the microbial community composition of the phyllosphere, the rhizosphere, and endophytic microbes. In this chapter, we discuss the relationship between the host and its symbiotic microbial community and the role of plant metabolites and root exudates such as organic acids, amino acids, sugars and antimicrobial compounds in shaping a specific rhizosphere community, attracting plant growth promoting rhizobacteria (PGPR) colonization on the plant roots and inhibiting or attracting soil-borne pathogens. In addition, we also review and introduce the functionality of plant symbiotic microbes for increasing the abiotic and biotic stress tolerance of the host. An understanding of the ecosystem function of plant and symbiotic microbes will guide efforts to improve agriculture practices.

Keywords: plant-microbe ecology, symbiotic microbial communities, interactions of host and symbiotic microbes

1. Introduction

Plants have recently been recognized as a metaorganism that possesses a distinct microbiome and close symbiotic relationships with associated microorganisms [1]. Plant ecology is affected by complex interactions with plant-associated microbes. The roles of both plant-associated microbes and the host in ecosystem function have been recognized, but the detailed mechanisms are unclear. Since plants are immobile, they have coevolved with microbes and acquired a number of mechanisms that modulate the outcome of their interactions [2]. Roots can continuously synthesize, accumulate, and secrete a wide range of compounds into the soil [3], which are known as the root exudates. Root exudates contain enzymes, water, H^+ ions, mucilage, and carbon-containing primary and secondary compounds [4, 5]. Campbell and Greaves [6, 7] observed that the density of microbes in the rhizosphere was 100 times greater than that in the bulk soil. Recent studies have showed that plant root exudates shape the soil bacterial community [8, 9]. According to a study by Ciccazzo [10], a plant species selects a specific rhizosphere bacterial community. The change in the microbial composition generates feedback on the plant relative performance that defines the long-term effects of the soil microbes on their coexistence with that plant species [11, 12]. The feedback can be of two types; positive plant-soil microbial feedback reinforces the spatial separation of the microbial communities [13], while negative feedback results in plant replacement, which necessitates recolonization of locally specific roots [14–16]. Systematic methods such as genome-wide association studies have enabled us to explore the relationships of plant loci and symbiotic communities in detail [17, 18]. How does the microbiome diversity and function potentially affect host plant performance? The presence of microbial hubs in plant microbiome networks plays an important role between a plant and its microbial community [19]. Plant growth–promoting rhizobacteria (PGPR) can produce a complex blend of volatile substances, which are distinct between bacterial species and other closely related species [20–22]. Some of these bacterial volatiles can stimulate plant growth [23, 24], suppress disease stimulating ISR [25], or antagonize phytopathogens [26, 27], nematodes, or insects [28, 29].

Worldwide crop production is affected by biotic and abiotic stress factors [30–32], which cause millions of dollars in losses. The beginning of the industrial revolution in 1750 and the human activities such as burning fossil fuels and deforestation have altered the global climate. An increase in carbon dioxide and temperature is speeding up the life cycle of grain and oilseed crops [33]. According to Lesk [34], extreme heat waves and droughts have reduced global harvests of cereals such as maize, wheat, and rice by 10% in a span of 50 years, which has become a grave concern of various governments. The impact of a warming climate on spring plant phenology is evident [35, 36]. A longer growing season may increase carbon uptake and potentially mitigate climate change [37, 38], leaf emergence [39], fruiting [40], and germination [41]. Abiotic stress factors include extreme temperature [42], drought, water logging, light, and salinity as major parameters that affect plant growth. Plant-associated microbes or PGPR were found to benefit plants by enhancing nutrient uptake, stimulating root, and shoot growth by producing indole acetic acid [43, 44], 1-aminocyclopropane-1-carboxylate (ACC) deaminases [45], solubilizing phosphate [46], and enhancing the uptake of nutrients from the environment [47, 48]. They also assist and enhance plant resistance to adverse environmental stresses such as drought, heavy metals, salts, and nutrient deficiency [49]. Biotic stress factors include interaction with other organisms and infection by pathogens or damage by insect pests, and some plant growth–promoting bacteria have been used as biocontrol agents against plant pathogens [50, 51]. This chapter explores the relationships between plant hosts and their symbiotic microbial communities. We discuss and review current reports of how the plant-associated microbial community might be shaped by the host and how the plant microbiome affects plant growth, productivity, and host survival in various symbiotic associations. This knowledge will guide efforts to improve agricultural practices and predict how environmental factors will affect the microbial community and plant diversity.

2. Plant community ecology and soil symbiont interactions

Plant ecology can be affected by global climate change in terms of above- and below-ground ecological diversity in a terrestrial ecosystem. The terrestrial plant community drives dynamic changes in the soil microbial ecology that may result in alterations in ecosystem function. The ecologist Dr. Peter Chesson postulated that stabilizing mechanisms are essential for maintenance of species diversity and coexistence [52]. Bever et al. have found evidence that microbially mediated positive and negative feedback might play a crucial role in the entire plant ecosystem and contribute to these mechanisms of plant-plant interactions [16]. Traditionally, competing plant species have been thought to have strong negative intraspecific interactions for the high overlap in resource usage [53]. However, success in finding a mechanism to explain the coexistence of competing plant species in maintaining local diversity [54]. Neglecting soil microbial community feedback might be one reason.

2.1. Soil resources and microbial interactions

Soil resources can govern the coexistence of plant species by resource partitioning and sharing. Studies have found root symbionts that increase the efficiency of nutrient uptake and allow the host to persist in a low nutrient environment, thereby directly contributing to the competitive exclusion of other plants [55]. Rhizosphere microbes can alter the availability of different forms of nitrogen or phosphorus in the soil and affect plant-plant interactions via the mediation of resource partitioning [56]. Soil resources can also be transferred by shared symbiotic fungi called common mycorrhizal networks (CMNs) [57]. In nature, different plant species commonly share the broadly specific mycorrhizal fungi. Simard and Durall demonstrated the direct transfer of resources from one plant to another via CMNs with labeled carbon, nitrogen, and phosphorus [57]. Plant community dynamics are driven by the microbial mediation of soil resource partitioning and sharing.

2.2. Host response to microbes and soil community feedback

The dynamic density and composition of the rhizospheric microbes can affect the coexistence of plant species via indirect feedback (i.e., the competition or inhibition of symbionts) in the plant population [58]. Ecologists have proposed three hypotheses to explain the mechanism that produces low diversity plant communities. The *empty niche* hypothesis suggests that novel symbionts inhabit the areas invaded by invasive plants [59]. These symbionts are more efficient at resource acquisition and preferentially associated with invasive plants than with other plants. The *degraded mutualist* hypothesis suggests that invasive plants and their symbionts inhibit the ability of the native symbiotic community to acquire resources, indirectly reducing the performance of native plants [60, 61] (**Figure 1**). Positive feedback might be exemplified by the enhanced growth and survival of exotic seedlings near native established symbionts [62, 63]. Plant monodominance, coexistence, and invasion ecology have high relation to symbiotic microbial interactions.

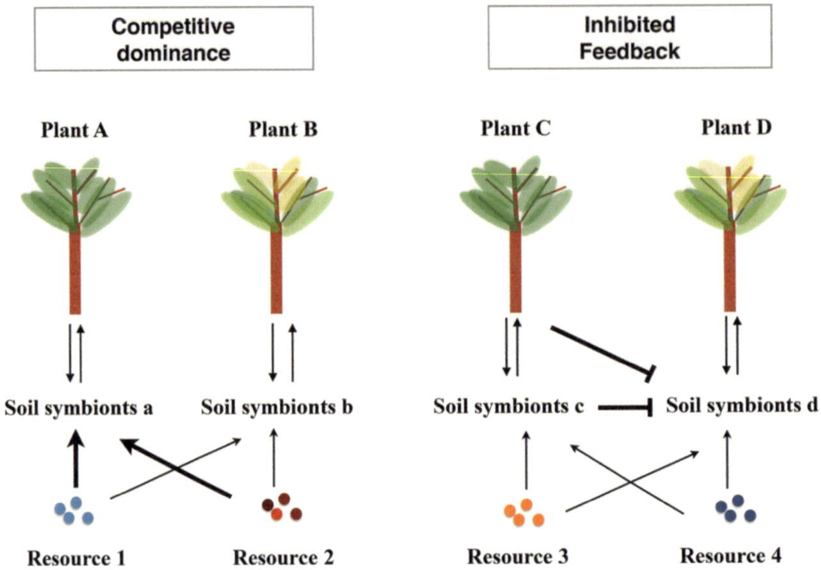

Figure 1. Plant soil community feedback in low diversity communities (modified after Bever et al. [16]).

3. Plants shape the microbial community

Microbial interactions play a crucial role in plant community ecology and performance. How do plants harbor unique microbial communities? How do plants shape a unique rhizosphere microbial community? These are the questions that must be addressed. Modern genomic technologies (e.g., high throughput sequencing) can provide clues to the answer. Lundberg et al. attempted to define the core *Arabidopsis thaliana* root microbiome [64]. They collected more than 600 *A. thaliana* plants and compared the bacterial communities using 16S rRNA gene sequencing. They observed that the root microbial communities of plants are sufficiently dependent on the host genotype to vary between inbred *A.* accessions. However, the mechanisms were not clear but included differences in the host physiology and immune responses.

3.1. Plant genes responsible for defense affect the variation of the microbial community

Several studies showed that plant genotype has a small but significant effect on the composition of the endophytic, rhizosphere, or phyllosphere microbial communities [17, 64–67]. A quantitative trait locus (QTL) analysis and a genome-wide association study (GWAS) were used to identify taxa linked to host genes in humans, mice, plants, and flies [17, 68–70]. A GWAS of the *A. thaliana* leaf microbial community suggested that the *A. thaliana* loci are responsible for defense and that cell wall integrity affected the composition of this community [17]. Furthermore, host genetic variation shaped species richness in the bacterial community. Horton et al. showed that defense response associated genes against pathogens shaped this microbial ecosystem. In Matthew's study, 196 accessions of *A. thaliana* were sown in a greenhouse and transferred to a field site. The field experiment data suggested that the plant tissue structure (i.e., the cell wall integrity) might affect the leaf microbial community. To understand the plant host genetic factors that affect the associated microbial population, Bodenhausen et al. used a candidate gene approach to investigate the host effects on the composition and abundance of the *A. thaliana* phyllosphere community [67]. A panel of 55 *A. thaliana* mutants with alterations in the cell wall, surface structure, defense signaling, secondary metabolism, and pathogen recognition was constructed to reveal the effect on the microbiota composition and/or abundance in a small number of single host mutations. The results showed that *lacs* and *pec1* mutants affected cuticle formation, which led to an increased bacterial abundance and community composition. Moreover, the ethylene signaling gene *ein2* was observed to be a host factor that modulated the community composition. Peiffer et al. also noted that approximately 19% of the interline variation in species richness could be attributed to the host genotype in different maize lines grown in the same geographical regions [65].

3.2. Role of root exudates in shaping rhizosphere microbial community

More than a century ago, Lorenz Hiltner defined the term "rhizosphere" as the soil compartment affected by plant roots [71]. Soil microbes are chemotactically attracted to plant root exudates, volatile organic carbon, and rhizodeposition, and then proliferate in this carbon-rich environment [72]. Plant root exudates differ between plant species, so differences in rhizosphere microbiomes of different plant species are expected [73]. More recent studies have provided strong evidence for plant species-specific microbiomes [74, 75]. Plants can also shape the microbial community via root exudates. Root exudates can be categorized as sugars, amino acids, organic acids, nucleotides, flavonoids, antimicrobial compounds, and enzymes [4, 73].

3.2.1. The types of root exudate

3.2.1.1. Organic acids and amino acids

The composition of root exudates from different cultivars affects the growth of soil-borne pathogens. The susceptible peanut cultivar Ganhua-5 (GH) and the mid-resistant cultivar Quanhua-7 (QH) were chosen for a root exudate analysis and evaluated for the responses of the soil-borne pathogens *Fusarium oxysporum* and *Fusarium solani* [76]. The contents of total amino acids, alanine, and sugars in the root exudate of susceptible cultivars were significantly higher than in the mid-resistant cultivar, whereas the contents of total phenolic acids, p-hydroxybenzoic acid, benzoic acid, and p-coumaric acid were significantly lower than in mid-resistant cultivars. These differences in the root exudate composition of susceptible and resistant cultivars might be assumed to regulate the resistance mechanism in the peanut rhizosphere. However, the spore germination and mycelial growth of the soil-borne pathogens *F. oxysporum* and *F. solani* were significantly enhanced by treatment with the root exudates from both the susceptible and mid-resistant cultivars compared with a control. If root exudates do not directly inhibit the growth of pathogens, the effects of other factors must be considered. A previous report showed that organic acids modulated the colonization and enhanced the biofilm formation of the root microbiome. Yuan et al. demonstrated that organic acids from banana root exudates facilitated the root colonization by *Bacillus amyloliquefaciens* [77]. Fumaric acid significantly induced biofilm formation, whereas malic acid evoked the greatest chemotactic response. The results showed that organic acids from banana root exudates played a crucial role in attracting and initiating PGPR colonization on the plant roots. Rice exudates that primarily contained the amino acid residues of histidine, proline, valine, alanine, and glycine, and the carbohydrates glucose, arabinose, mannose, galactose, and glucuronic acid may induce a higher chemotactic response by the endophytic bacteria *Corynebacterium flavescens* and *Bacillus pumilus* [78].

3.2.1.2. Sugars

The amount of sugar secretion might affect infection by plant pathogens. Gou et al. showed that the *Arabidopsis* vacuolar sugar transporter SWEET2 limited the Glc-derived carbon efflux from roots and inhibited *Pythium* infection [79]. They proposed that the expression of SWEET2 modulated sugar secretion, limiting the carbon loss to the rhizosphere. The reduction of available substrates in the rhizosphere contributed to the resistance to *Pythium*.

3.2.1.3. Antimicrobial compounds

Root exudates can also participate in belowground plant defense. Low-molecular-weight antimicrobial chemicals can be divided into "phytoanticipins" and "phytoalexins" [80]. Phytoanticipins are defensive compounds that are constitutively produced and released by the plant root prior to a biotic stress such as pathogen infection. In a recent study, *Arabidopsis* roots deficient in diterpene rhizathalene A production were found to be more susceptible to insect herbivory [81]. Therefore, rhizathalene A was considered as a part of a constitutive direct defense system of the roots. Phytoalexins were defined as inducible defensive compounds that are not detected in healthy plants [80]. Five phenylpropanoid root-derived aromatic root exudates were induced by the attack of the soil-borne pathogen *Fusarium graminearum* and exhibited antifungal activity [82]. In general, root-secreted terpenoid and phenolic defensive compounds have strong antibacterial and antifungal activity [83, 84]. The largest class of plant defensive chemicals above- and below ground is terpenoids. Nonvolatile terpenoids can be secreted into the rhizosphere [85], and volatile organic compounds (VOCs) can be emitted from the roots as plant defensive compounds. Phenylpropanoids are a group of plant defensive phenolic root exudates. After a Fusarium graminearum infection, barley rapidly accumulated and secreted phenylpropanoids, which are cinnamic acid derivatives to resist a fungal attack [82]. Phenolic root exudates not only have antimicrobial activity but also beneficially attract soil-borne microorganisms that affect the native soil microbial community [86]. We have found that the same chemical compound, for example, the amino acid canavanine, can stimulate a specific group of microbes but suppress many other soil microbes. Plants can shape the specific rhizosphere microbial community via root exudates.

3.2.2. *Environmental factors effects on root exudates*

Plants with different genotypes produce root exudates with different compositions. Abiotic and biotic factors also affect root exudates. Physico-chemical soil properties such as nutrient availability, organic matter content, pH, structure, and texture can affect the availability of root exudates and microbial recruitment by the plant roots. Some biotic factors such as soil microbial secondary metabolism can also affect the exudates.

3.2.2.1. *Temperature*

Since the onset of climate change and global warming, the resultant extreme heat and cold have affected the harvest of several crops. To elucidate the effects of temperature on root exudates, Husain and McKeen grew strawberry plants at 5–10°C and compared them with plants grown at 20–30°C. They found more amino acids in exudates in plants grown at a low soil temperature that markedly affected the pathogenicity of *Rhizoctonia fragariae* [87]. Pramanik demonstrated that in Japanese cucumber grown hydroponically in a growth chamber at high and low temperatures, the organic acid content increased with the elevation of temperature, and some of the compounds identified significantly inhibited plant root growth and/or germination of cucumber.

3.2.2.2. *Soil moisture*

Flood and drought have reduced global cereal harvests. Several reports have demonstrated that the soil moisture affects the release of root exudates. The temporarily wilting of plants increased the release of amino acids from the plant roots, which might be related to the incidence of pathogens in the rhizosphere [88]. Plants such as peas, soybeans, wheat, barley, and tomatoes were grown in normal moist sand and dried, remoistened sand for the liberation of amino acids. The total amount of amino nitrogen in the temporarily dried sand was many times higher than in the normal moist sand.

3.2.2.3. *Soil pH and nutrition*

The soil pH status and the availability of nutrients such as carbon, nitrogen, and phosphate have been found to affect the release of plant root exudates and the creation of specific chemical niches in the soil, as well as the abundance of plant pathogens and beneficial microbes [89–91]. Bowen first demonstrated a marked effect of nutrient status on the exudation of amides and amino acids from roots of *Pinus radiate* seedlings [89]. The results indicated a doubling of amides/amino acids in exudates from phosphate deficient plants. Toljander et al. analyzed the community of arbuscular mycorrhizal (AM) fungi in maize in a long-term fertilization trial and indicated the composition of AM fungi and bacteria was significantly affected and correlated with changes in pH, phosphate, and the soil carbon content [90]. Dumbrell et al. surveyed the AM fungal community of 425 individual plants from 28 plant species. The results showed the strong support for the hypothesis that niche differentiation was based on the structuring of the AM fungal community by soil pH [91]. Root secretion of phenolics was induced in Fe-deficient soil and altered the microbial community in the rhizosphere [92].

3.2.2.4. Microorganisms

Soil microorganisms play a crucial role in plant growth and plant exudates. Microorganisms can affect exudation by affecting the permeability of root cells and root metabolism. Microorganisms can also absorb certain compounds in root exudates and excrete other compounds. Soil microbes can produce secondary metabolites that affect plant signaling and metabolism and can be considered as a "plant secondary genome" that provides plant hosts with microbe-derived compounds [93]. Some microbes and also some antibiotics (e.g., penicillin and polymyxin) increased the exudation of organic materials, altered cell permeability, and increased leakage [94, 95]. Soil microbes can also induce the exudation of phenolic compounds for enhancing plant Fe absorption in low-Fe availability soil [96].

4. Microbial community diversity and plant performances

4.1. Variation of microbial community in plant life cycle

Plant and rhizosphere microbial diversity varies throughout the plant life cycle. The factors influencing the composition and diversity of the microbial community can be classified as four processes: dispersal, drift, speciation, and selection [97]. For seed plants, the life cycle begins with a seed. Seed dispersal is an important ecological process. Seeds carry associated microbes that originate from their parent and the environment, thereby increasing the microbial diversity in a new environment. Recent studies have suggested that bacterial seed coatings can protect against pathogens [98]. Microbial seed epiphytes have an advantage over soil bacteria during plant colonization. Seed coating methods are a major area of research, and numerous patents have been filed (i.e., approximately 4000 results were found by a Google patent search for the key word "microbial seed coating" [99]. After seed dispersal, during seed germination, seed-bone microbes might gain a competitive advantage over other microbes to colonize after germination, and opportunistic microbes from the surrounding soil might have access to a novel niche as the plant develops. Microbial diversity and the community dynamically change throughout the plant life cycle.

4.2. Networking of plant-microbes (hub and edge microbes)

Plant microbiota forms a complex network. A wide range of studies has demonstrated that plant-associated microbes live either inside plant tissue or on the surface of plant organs such as the leaves and roots [100, 101]. Agler et al. characterized the microbiome of *A. thaliana* leaves [102]. Field experiments showed that both plant genotype and abiotic factors affected the microbiome composition. In addition, they observed that specific species (e.g., the plant pathogen *Albugo* and the fungus *Dioszegia*) significantly affected the microbial community structure. Agler used the term "microbial hubs" for the presence of these specific species, which were strongly interconnected with other species in the microbial network of the plants.

Microbial hubs might be responsible for mediating defense signals among plants and the effectiveness of biological control agents [19]. The term "keystone species" has been proposed for the presence of a kind of hub species that would be a determinant of colonization of widely microbial taxa. These microbial hubs and keystone species have a large impact on plant performance. A number of hypothetical relationships between plant performance and microbial diversity and composition have been proposed [19].

Microbial hubs might indirectly affect other taxa by changing host performance, response, or metabolites without directly interacting with other microbes. How can the microbial hubs be identified and how can the interaction of plants and microbes be understood? The requirement of new techniques to analyze whether a microbe has successfully entered a plant, and the observation of changes in the genotypic and phenotypic expression will be an added advantage in the study of plant-microbe interactions.

5. Plant growth–promoting microbes

The soil constitutes a pool of microscopic life forms including bacteria, fungi, actinomycetes, protozoa, and algae, and of these, bacteria are by far the most common. The highest numbers of bacteria are found in the rhizosphere, the region around the plant roots, as differentiated from the bulk soil [103]. Regardless of the concentration of bacteria in the soil, the bacteria may affect a plant in one of three ways. From the perspective of the plant, the interaction between the soil bacteria and a plant may be beneficial, harmful, or neutral [104]. Plant growth-promoting bacteria (PGPB) include those that are free living, those that form specific symbiotic relationships with plants (e.g., *Rhizobia* spp. and *Frankia* spp.), bacterial endophytes that can colonize some, or a portion of a plant's interior tissues and cyanobacteria. PGPB can promote plant growth directly by facilitating the acquisition of compounds or modulating plant hormone levels and indirectly by reducing the inhibitory effect of pathogenicity and plant growth by acting as biocontrol agents [105].

5.1. PGPB and abiotic stress

In nature, all living organisms are affected by environmental factors such as abiotic stress. Some plants have internal mechanisms to cope up with such stress, while others overcome. Abiotic stress factors include water deficit, excessive water, extreme temperatures, and salinity. The association of PGPB with certain plants can help the plants combat certain abiotic stresses and prevent the plants from dying. In the past decade, bacteria belonging to different genera including *Rhizobium, Bacillus, Pseudomonas, Pantoea, Paenibacillus, Burkholderia, Achromobacter, Azospirillum, Microbacterium, Methylobacterium, Variovorax,* and *Enterobacter* have been reported to endow host plants under different abiotic stress environments [106].

5.1.1. Cold stress

Maize plants exposed to low temperatures show reduced shoot and root growth that has been attributed to severe oxidative damage induced by cold stress [107, 108]. Treatment with *Pseudomonas* sp. DSMZ 13134, *B. amyloliquefaciens* subsp. *plantarum, Bacillus* simplex strain

R41 with micronutrients (Zn/Mn), or seaweed extracts proved to be beneficial cold stress protectant [109]. Inoculation of tomato seeds with plant growth–promoting psychrotolerant bacteria from the genera *Arthrobacter, Flavobacterium, Flavimonas, Pedobacter,* and *Pseudomonas* significantly improved plant height, root length, and membrane damage in leaf tissues as evidenced by electrolyte leakage and the malondialdehyde content [110]. A cold-tolerant PGPB *Methylobacterium phyllosphaerae* strain IARI-HHS2-67, isolated using a leaf imprinting method from phyllosphere of wheat (*Triticum aestivum* L.), showed improved survival, growth, and nutrient uptake compared to a noninoculated control at 60 days under low-temperature conditions [111]. The chilling resistance of grapevine plantlets was enhanced when inoculated with a plant growth–promoting rhizobacteria, *Burkholderia phytofirmans* strain PsJN. The root growth increased by 11.8- and 10.7-fold at 26 and 4°C, respectively, and plantlet biomass increased by 6- and 2.2-fold at 26 and 4°C, respectively [112].

5.1.2. Heat stress

The effects of global warming in recent years can be felt with the increase in global temperature. A thermo tolerant plant growth–promoting *Pseudomonas putida* strain AKMP7 was proven to be beneficial for the growth of wheat (*Triticum* spp.) under heat stress [113]. The bacterium significantly increased the root and shoot length and dry biomass of wheat as compared to uninoculated plants. Inoculation improved the level of cellular metabolites and reduced the activity of several antioxidant enzymes and membrane injury. Sorghum seedlings showed enhanced tolerance to increased temperature with the association of *Pseudomonas* sp. strain AKM-P6 [114]. Inoculation induced the biosynthesis of high-molecular-weight proteins in the leaves at elevated temperatures, reduced membrane injury, and improved the levels of cellular metabolites such as proline, chlorophyll, sugars, amino acids, and proteins.

5.1.3. Salinity

Approximately 20–50% of crop yields are lost to drought and high soil salinity [115]. The United Nations Population Fund estimates that the global human population may well reach 10 billion by 2050 (www.unfpa.org). Crop plants are very sensitive to soil salinity, and it is one of the harshest environmental factors that limits the productivity of crops. Plant-microbe associations have been found to be beneficial against abiotic salt stress in *Zea mays* upon coinoculation with Rhizobium, while Pseudomonas was correlated with decreased electrolyte leakage and the maintenance of leaf water content [116]. Salinity resistant *Pseudomonas fluorescens, P. aeruginosa,* and *P. stutzeri* ameliorated sodium chloride stress in tomato plants, and an increase in roots and length were observed [117]. Jha et al. demonstrated that the endophytic bacteria *Pseudomonas pseudoalcaligenes* induced the accumulation of higher concentrations of glycine betain-like compounds, leading to improved salinity stress tolerance in rice [118]. *Dietzia natronolimnaea,* a plant growth–promoting rhizobacteria, was seen to modulate a stress response gene, which led to the protection of wheat from salinity stress [119]. *Staphylococcus saprophyticus* ST1 and *Oceanobacillus profundus* Pmt2 inoculants were able to produce a biofilm and an extracellular EPS, thus helping *Lens esculenta* Var. Masoor-93 to cope with salt stress [120]. Salt-stressed *Arabidopsis* plants treated with volatile organic compounds (VOCs) from *B. amyloliquefaciens* GB03 showed higher biomass production and less Na^{+} accumulation compared to salt-stressed plants without VOC treatment [121].

5.1.4. Water stress resistance

Water scarcity constrains plant productivity, and more crop productivity is lost due to water scarcity than any other abiotic stresses [122]. *Achromobacter piechaudii* ARV8 reduced the production of ethylene by tomato seedlings following water stress, and ARV8 did not affect the reduction of the relative water content during water deprivation. ARV8 significantly improved the recovery of plants when watering was resumed [123]. Water stress resistance was enhanced in green gram when treated with *P. fluorescens* Pf1 compared to untreated plants. *P. fluorescens* Pf1 was also found to produce the enzyme catalase under stress conditions, which helped to detoxify the compounds accumulated in green gram during adverse conditions [124].

Heavy metals are defined as metals with a density higher than 5 g/cm^3 [125]. Heavy metals cause a significant decrease in plant growth and protein content at high concentrations. The most common heavy metal contaminants are Hg, Cd, Cr, Cu, Pb, and Zn [126]. All of these elements are toxic to crop plants at high tissue concentrations. Heavy metal toxicity in plants leads to the production of reactive oxygen species that block essential functional groups of biomolecules. This reaction has been noted in Hg and Cd toxicity and causes oxidative injury in plants. Increasing concentrations of Hg (5–20 mg/kg soil) in tomato plant showed deleterious effects on survival percentage, germination, flowering, pollen viability, and reduced plant height. *P. putida* enhanced the Cd uptake potential of *Eruca sativa* and favored healthy growth under Cd stress by increasing the shoot length up to 27%, the root length up to 32%, the wet weight up to 40%, the dry weight up to 22%, and the chlorophyll content up to 26% [127]. Canola seeds inoculated with *Kluyvera ascorbata* SUD165 and grown under gnotobiotic conditions in the presence of high concentrations of nickel chloride were partially protected against nickel toxicity because the bacteria could lower ethylene-induced stress due to nickel toxicity [128]. *Photobacterium halotolerans* MELD1 facilitated the phytoprotection of *Vigna Unguiculata Sesquipedalis* against Hg at a concentration of 25 ppm, thus increasing productivity as well as reducing the translocation of Hg to the bean pods [121]. A plant-microbe phytoremediation system was created with the combination of vetiver grass and the functional endophytic bacterium *Achromobacter xylosoxidans* F3B for the removal of toluene in Ho et al. [129]. It was observed that *A. xylosoxidans* F3B improved the degradation of toluene in vetiver, resulting in a decrease in phytotoxicity and a 30% reduction of evapotranspiration through the leaves. Another study conducted by Ho et al. [130] observed that when *A. xylosoxidans* strain F3B was inoculated in *A. thaliana*, it helped the plant tolerate a lethal concentration of catechol and phenol and enhanced the phytoremediation and phytoprotection of the plant.

5.2. PGPB against biotic stress

Plants must withstand adverse abiotic and biotic stresses when they are sessile (**Figure. 2**). Biotic stress in plants mainly includes damage caused by other living organisms such as insects, bacteria, fungi, nematodes, viruses, viroids, and protists. Biotic stress by PGPR can affect plant growth in two different ways; by the direct promotion of plant growth by the production of phytohormones or by facilitating the uptake of certain nutrients [45]. The indirect promotion of plant growth occurs when PGPR lessens or prevents the deleterious effects of phytopathogens. *P. fluorescens* produces 2,4-diacetyl phloroglucinol, which inhibits the growth of phytopathogenic fungi [131]. Extracellular chitinase and laminarinase were

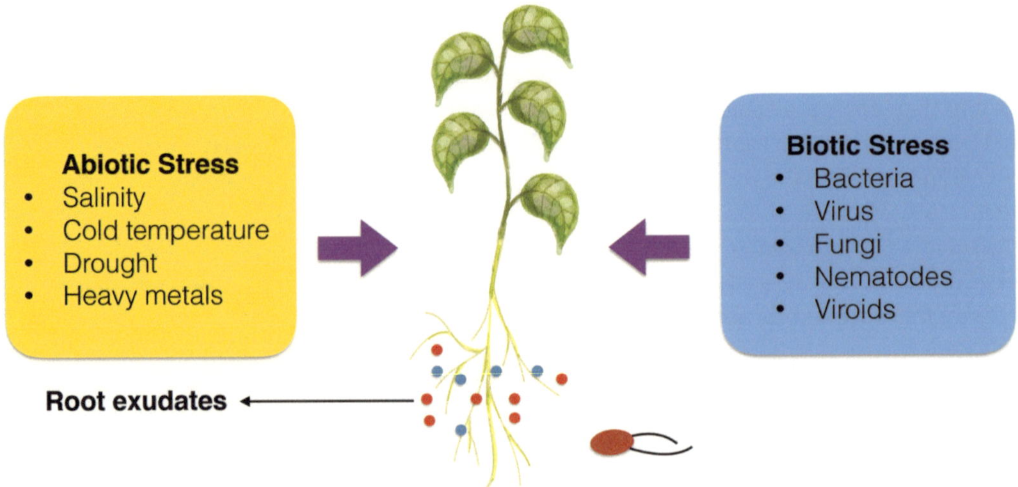

Figure 2. Factors affecting plant-microbe interactions.

produced by *P. stutzeri,* which caused the lysis of mycelia of *F. solani,* which causes root rot [132]. The endophytic *B. cenocepacia* 869T2 decreased the disease incidence of Fusarium wilt in treated banana plants to 3.4%, compared to 24.5% in noninoculated plants infected in a field test during a 7-month period [50]. The antibiotic Pyrrolnitrin, produced by *P. fluorescens* BL915 could prevent the damage from *Rhizoctonia solani* during the damping off of cotton plants [133].

Van Peer et al. [134] described a mechanism called "Induced Systemic Resistance" in carnation plants that were systematically protected by *P. fluorescens* strain WCS417r against *F. oxysporum* f. sp. *Dianthi* and by Ardebili et al. in tomato plants, in which *P. fluorescens* CHA0 protected against *F. oxysporum f.* sp. Lycopersici acted as a bio agent that induced resistance in tomato [135]. *B. amyloliquefaciens* strain FZB42, a plant root colonizing isolate, was seen to have the ability to stimulate plant growth and suppress plant pathogens [136]. In another study, endophytes were seen to protect cucumber plants against cucumber anthracnose induced by *P. fluorescens* strain 89B-61 [137] and *Achromobacter* sp. F2feb.44. *Streptomyces* sp. Zapt10, and *Bacillus licheniformis* AE6 were exploited to induce systemic resistance in cucumber against the foliar disease of downy mildew caused by the phytopathogen *Pseudoperonospora cubensis,* which enhanced yield [138].

Pest management has become an issue over time because more and more pests are becoming immune to pesticides. The global pesticide market is growing at a pace of 3.6% per year and is valued around $47 billion [139]. Development of entomopathogenic bacteria for pest management has been a new approach to handle resilient pests. Species belonging to the genera *Aschersonia, Agerata, Verticillium, Sphaerostilbe, Podonectria, Myriangium, Hirsutella,* and *Metarhizium* [140] are fungal species involved in the biological control of pests. *Bacillus thuringiensis* is the most well-known *Bacillus* species on which the efforts of the scientific community and industry have been focused [141]. *Brevibacillus laterosporus* Laubach action has been reported to be effective against insects such as Coleoptera [142], Lepidoptera [143], nematodes [144, 145], and phytopathogenic fungi [146].

6. Future development of application

6.1. Techniques used to understand plant microbe interaction

Plants have been an integral part of our diet since humans began to farm and were no longer nomads. Since then, the world has faced a continuous challenge to feed the ever-growing population. The excess use of fertilizer is causing eutrophication [147], and genetic engineering of plants is an expensive and tedious process. The role of microbes in plant-microbe interactions has been studied in detail during the past decade [148, 149]. The results of extensive studies suggest that the exploitation of beneficial microbes is a better strategy in the long run to increase crop yield, which will play an important role in disease dissemination and control [150]. Three types of plant-microbe interactions have been studied–symbiosis between plants and mycorrhizae [151], between plants and rhizobacteria [152], and pathogenesis [153]. Omics technologies such as transcriptomics, proteomics, and metabolomics coupled with bioinformatics have been extensively applied in plant abiotic stress studies [154]. The proteomics approach has been largely adopted to investigate the protein profiles in plants in response to abiotic stresses that might lead to the development of new strategies for improving stress tolerance [155]. Microbial metabolomics is another technique that has been used to study the set of metabolites present in microbial communities [156]. A rhizosphere metabolomics–driven approach was used in the study of plant-microbe interactions for the removal of polychlorinated biphenyls as demonstrated by Lee et al. [157].

Pathogens and their emergence have been a great threat to food security, agricultural practices, and the conservation of food species, and it has become a significant task to understand the emergence of new pathogens and their role [158]. In the past, plant pathosystems were best studied one gene at a time or one protein at a time; however, the genomic era marked the beginning of the in-depth study of plant-pathogen associations [159]. Commonly known pathogen genomes have been sequenced, and the analysis of these sequences has revealed the forces that have shaped pathogen evolution and has brought to light the unexpected aspects of pathogen biology [160, 161]. The genome scale reconstruction model (GSRM) is based on metabolic reconstructions on a genomic scale for the analysis and interpretation of metabolite concentrations under specific conditions [162]. GSRM has been successfully developed for many organisms including plants, bacteria, fungi, and animals [163–165].

Various techniques to study plant-microbe interactions are sequencing, chromatography, mass spectrometry, phospholipid fatty acid (PLFA), microscopy, Fourier transform infrared spectroscopy (FTIR), nuclear magnetic resonance, and real-time PCR (RT-PCR) [166]. Further advances in the postgenomic era will pave the way for a better understanding of the interactions of endophytes, plant-pathogen, and plant immunity. Genomic tools to understand major units of the host-microbiota ecology are shown in **Figure 3**.

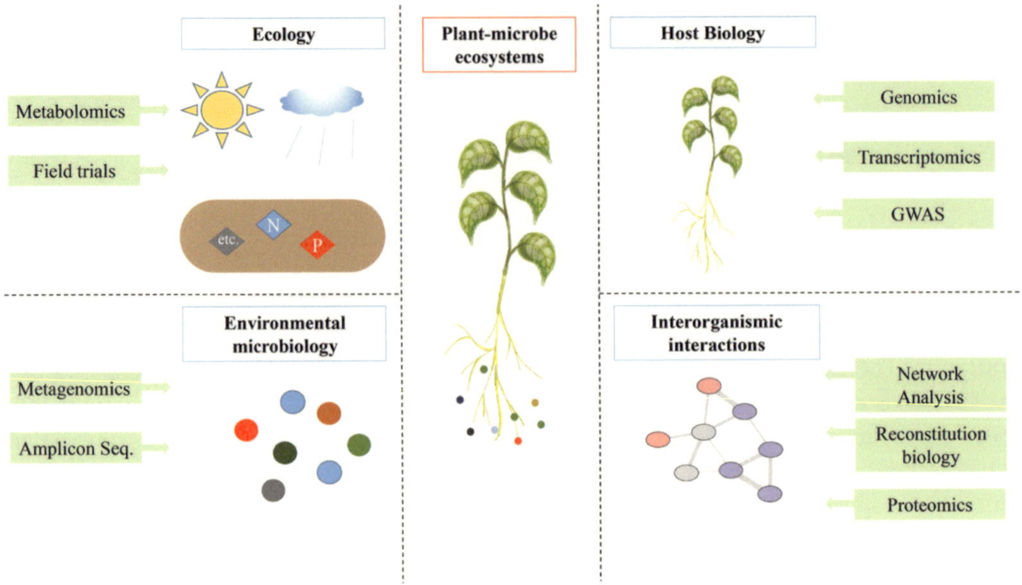

Figure 3. Genomic tools to understand major units of the plant-microbe ecological system. Colored boxes indicate technique tools that can be used to characterize key factors for the corresponding unit. (modified after Kroll et al. [167]).

7. Conclusions

In the face of pressure from climate change, contaminated environments, and crop pathogens, agricultural material and food production are currently at risk. Plant-associated microorganisms have important consequences for host health and performance. However, efforts to utilize beneficial microbes in the field have failed to consistently improve crops. The current understanding of interactions between plants and symbiotic microbial communities, the ecological consequences of plant-associated microorganisms and plant-microbial metabolic dynamics are limited. The advent of genomic approaches has helped a great deal in the understanding of the plant-bacterial interactions, but genomic approaches are still insufficient to clearly explain the interactions between plants and pathogens [168]. Approaches using metagenomics and amplicon sequencing coupled with other omics technologies [169] and the development of databases (PHI-based) [170], and metabolomics have enhanced our understanding of plant-bacterial interactions. Plant-microbe ecological communities are affected by plant genotype and environmental factors. The difference between genotypes causes different physiological and immune responses and leads to host-specific microbial communities. Plant root exudates (i.e., sugars, amino acids, organic acids, nucleotides, flavonoids, antimicrobial compounds, and enzymes) shape specific communities, attract plant growth-promoting colonization, and pathogen infections.

In the complex microbial community, we found that specific species could significantly affect the microbial community structure. Via a systems' framework of microbial network analysis, we could identify the "hub species" and "milestone species", which are candidate microbial assemblages for disease management. Network models of plant-associated microbiomes provide new opportunities for enhancing disease management.

7.1. Future perspective

The understanding of plant-microbe and microbe-microbe interactions will provide a helpful future perspective as a modulating microbiome for minimizing disease incidence and enhancing gross plant productivity. Further, beneficial plant-associated bacteria could act as counterparts against pathogens within the microbial ecosystem, as well as stabilize the ecosystem, enhance biodiversity, avoid pathogen outbreaks, and increase plant productivity. A well-studied plant-microbe partnership in the future will also help increase crop productivity at little expense and could, in turn, lead to another "Green Revolution".

Acknowledgements

The authors are grateful to the National Chung Hsing University, College of Life Science. The research work has been done with the financial support by Council of Agriculture, Executive Yuan (grant 104AS-16.3.1-ST-aB) and the Ministry of Science and Technology of Taiwan (grant MOST 104-2321-B-001-060-MY3).

Author details

Ying-Ning Ho[1], Dony Chacko Mathew[2] and Chieh-Chen Huang[2]*

*Address all correspondence to: cchuang@nchu.edu.tw

1 Agricultural Biotechnology Research Center, Academia Sinica, Taipei, Taiwan

2 Department of Life Sciences, National Chung Hsing University, Taichung, Taiwan

References

[1] Mendes R, Garbeva P, Raaijmakers JM. The rhizosphere microbiome: Significance of plant beneficial, plant pathogenic, and human pathogenic microorganisms. FEMS Microbiology Reviews. 2013;**37**(5):634-663

[2] Oldroyd GE. Speak, friend, and enter: Signalling systems that promote beneficial symbiotic associations in plants. Nature Reviews Microbiology. 2013;**11**(4):252-263

[3] Jones JD, Dangl JL. The plant immune system. Nature. 2006;**444**(7117):323-329

[4] Baetz U, Martinoia E. Root exudates: The hidden part of plant defense. Trends in Plant Science. 2014;**19**(2):90-98

[5] Wen F, VanEtten HD, Tsaprailis G, Hawes MC. Extracellular proteins in pea root tip and border cell exudates. Plant Physiology. 2007;**143**(2):773-783

[6] Faure D, Vereecke D, Leveau JH. Molecular communication in the rhizosphere. Plant and Soil. 2009;**321**(1-2):279-303

[7] Campbell R, Greaves M. Anatomy and community structure of the rhizosphere. The Rhizosphere. 1990:11-34

[8] Haichar FZ, Marol C, Berge O, Rangel-Castro JI, Prosser JI, Balesdent J, et al. Plant host habitat and root exudates shape soil bacterial community structure. The ISME Journal. 2008;**2**(12):1221-1230

[9] Lareen A, Burton F, Schäfer P. Plant root-microbe communication in shaping root microbiomes. Plant Molecular Biology. 2016;**90**(6):575-587

[10] Ciccazzo S, Esposito A, Rolli E, Zerbe S, Daffonchio D, Brusetti L. Different pioneer plant species select specific rhizosphere bacterial communities in a high mountain environment. SpringerPlus. 2014;**3**(1):391

[11] Bever JD. Soil community feedback and the coexistence of competitors: Conceptual frameworks and empirical tests. New Phytologist. 2003;**157**(3):465-473

[12] Bever JD, Westover KM, Antonovics J. Incorporating the soil community into plant population dynamics: The utility of the feedback approach. Journal of Ecology. 1997;**85**(5):561-573

[13] Alexander I, Lee S. Mycorrhizas and Ecosystem Processes in Tropical Rain Forest: Implications for Diversity. Biotic Interactions in the Tropics: Their Role in the Maintenance of Species Diversity. Cambridge University Press: New York. 2005. pp. 165-203

[14] Waller F, Mueller MJ, Pedrotti L. Piriformospora indica root colonization triggers local and systemic root responses and inhibits secondary colonization of distal roots. PloS one. 2013;**8**(7):e69352

[15] Wagg C, Boller B, Schneider S, Widmer F, van der Heijden MG. Intraspecific and intergenerational differences in plant–soil feedbacks. Oikos. 2015;**124**(8):994-1004

[16] Bever JD, Dickie IA, Facelli E, Facelli JM, Klironomos J, Moora M, et al. Rooting theories of plant community ecology in microbial interactions. Trends in Ecology & Evolution. 2010;**25**(8):468-478

[17] Horton MW, Bodenhausen N, Beilsmith K, Meng D, Muegge BD, Subramanian S, et al. Genome-wide association study of *Arabidopsis thaliana*'s leaf microbial community. Nature Communications. 2014;**5**:5320

[18] Herrera Paredes S, Lebeis SL. Giving back to the community: microbial mechanisms of plant–soil interactions. Functional Ecology. 2016;**30**(7):1043-1052

[19] van der Heijden MG, Hartmann M. Networking in the plant microbiome. PLoS Biology. 2016;**14**(2):e1002378

[20] Garbeva P, Hordijk C, Gerards S, de Boer W. Volatile-mediated interactions between phylogenetically different soil bacteria. Frontiers in Microbiology. 2014;**5**:285-290

[21] Garbeva P, Hordijk C, Gerards S, De Boer W. Volatiles produced by the mycophagous soil bacterium Collimonas. FEMS Microbiology Ecology. 2014;**87**(3):639-649

[22] Groenhagen U, Baumgartner R, Bailly A, Gardiner A, Eberl L, Schulz S, et al. Production of bioactive volatiles by different *Burkholderia ambifaria* strains. Journal of Chemical Ecology. 2013;**39**(7):892-906

[23] Ryu C-M, Farag MA, Hu C-H, Reddy MS, Wei H-X, Paré PW, et al. Bacterial volatiles promote growth in Arabidopsis. Proceedings of the National Academy of Sciences. 2003;**100**(8):4927-4932

[24] Peñuelas J, Asensio D, Tholl D, Wenke K, Rosenkranz M, Piechulla B, et al. Biogenic volatile emissions from the soil. Plant, Cell & Environment. 2014;**37**(8):1866-1891

[25] Yi H-S, Yang JW, Ryu C-M. ISR meets SAR outside: Additive action of the endophyte *Bacillus pumilus* INR7 and the chemical inducer, benzothiadiazole, on induced resistance against bacterial spot in field-grown pepper. Frontiers in Plant Science. 2013;**4**:122

[26] Pineda A, Dicke M, Pieterse CM, Pozo MJ. Beneficial microbes in a changing environment: Are they always helping plants to deal with insects? Functional Ecology. 2013;**27**(3):574-586

[27] Kai M, Haustein M, Molina F, Petri A, Scholz B, Piechulla B. Bacterial volatiles and their action potential. Applied Microbiology and Biotechnology. 2009;**81**(6):1001-1012

[28] Boyer JS. Plant productivity and environment. Science. 1982;**218**(4571):443-448

[29] Bray EA, Bailey-Serres J, Weretilnyk E. Responses to abiotic stresses. Biochemistry and Molecular Biology of Plants. 2000;**1158**:e1203

[30] Suzuki N, Rivero RM, Shulaev V, Blumwald E, Mittler R. Abiotic and biotic stress combinations. New Phytologist. 2014;**203**(1):32-43

[31] Hua J. Modulation of plant immunity by light, circadian rhythm, and temperature. Current Opinion in Plant Biology. 2013;**16**(4):406-413

[32] Sachdev DP, Chaudhari HG, Kasture VM, Dhavale DD, Chopade BA. Isolation and characterization of indole acetic acid (IAA) producing *Klebsiella pneumoniae* strains from rhizosphere of wheat (*Triticum aestivum*) and their effect on plant growth. Indian Journal of Experimental Biology. 2009;**47**:993-1000

[33] Backlund P, Janetos A, Schimel D. The effects of climate change on agriculture, land resources, water resources, and biodiversity in the United States. Washington, DC: United States Climate Change Science Program. 2008

[34] Lesk C, Rowhani P, Ramankutty N. Influence of extreme weather disasters on global crop production. Nature. 2016;**529**(7584):84-87

[35] Ma T, Zhou C. Climate-associated changes in spring plant phenology in China. International Journal of Biometeorology. 2012;**56**(2):269-275

[36] Allstadt AJ, Vavrus SJ, Heglund PJ, Pidgeon AM, Thogmartin WE, Radeloff VC. Spring plant phenology and false springs in the conterminous US during the 21st century. Environmental Research Letters. 2015;**10**(10):104008

[37] Black T, Chen W, Barr A, Arain M, Chen Z, Nesic Z, et al. Increased carbon sequestration by a boreal deciduous forest in years with a warm spring. Geophysical Research Letters. 2000;**27**(9):1271-1274

[38] Dragoni D, Schmid HP, Wayson CA, Potter H, Grimmond CSB, Randolph JC. Evidence of increased net ecosystem productivity associated with a longer vegetated season in a deciduous forest in south-central Indiana, USA. Global Change Biology. 2011;**17**(2):886-897

[39] Jeong SJ, Ho CH, Gim HJ, Brown ME. Phenology shifts at start vs. end of growing season in temperate vegetation over the Northern Hemisphere for the period 1982-2008. Global Change Biology. 2011;**17**(7):2385-2399

[40] Xia J, Wan S. Independent effects of warming and nitrogen addition on plant phenology in the Inner Mongolian steppe. Annals of Botany. 2013;**111**(6):1207-1217

[41] Orsenigo S, Abeli T, Rossi G, Bonasoni P, Pasquaretta C, Gandini M, et al. Effects of autumn and spring heat waves on seed germination of high mountain plants. PLoS One. 2015;**10**(7):e0133626

[42] Majeed A, Abbasi MK, Hameed S, Imran A, Rahim N. Isolation and characterization of plant growth-promoting rhizobacteria from wheat rhizosphere and their effect on plant growth promotion. Frontiers in Microbiology. 2015;**6**:198

[43] Honma M, Shimomura T. Metabolism of 1-aminocyclopropane-1-carboxylic acid. Agricultural and Biological Chemistry. 1978;**42**(10):1825-1831

[44] Otieno N, Lally RD, Kiwanuka S, Lloyd A, Ryan D, Germaine KJ, et al. Plant growth promotion induced by phosphate solubilizing endophytic Pseudomonas isolates. Frontiers in Microbiology. 2015;**6**:745

[45] Glick BR. The enhancement of plant growth by free-living bacteria. Canadian Journal of Microbiology. 1995;**41**(2):109-117

[46] Glick BR, Patten CL, Holguin G, Penrose D. Biochemical and genetic mechanisms used by plant growth promoting bacteria. Imperial College Press, London, UK. 1999

[47] Egamberdieva D, Jabborova D, Hashem A. Pseudomonas induces salinity tolerance in cotton (*Gossypium hirsutum*) and resistance to Fusarium root rot through the modulation of indole-3-acetic acid. Saudi Journal of Biological Sciences. 2015;**22**(6):773-779

[48] Bano A, Fatima M. Salt tolerance in Zea mays (L). following inoculation with Rhizobium and Pseudomonas. Biology and Fertility of Soils. 2009;**45**(4):405-413

[49] Zhang H, Kim M-S, Sun Y, Dowd SE, Shi H, Paré PW. Soil bacteria confer plant salt tolerance by tissue-specific regulation of the sodium transporter HKT1. Molecular Plant-Microbe Interactions. 2008;**21**(6):737-744

[50] Ho Y-N, Chiang H-M, Chao C-P, Su C-C, Hsu H-F, Guo C-t, et al. In planta biocontrol of soilborne Fusarium wilt of banana through a plant endophytic bacterium, *Burkholderia cenocepacia* 869T2. Plant and Soil. 2015;**387**(1-2):295-306

[51] Prashar P, Kapoor N, Sachdeva S. Biocontrol of plant pathogens using plant growth promoting bacteria. In: Sustainable Agriculture Reviews: Springer Netherlands. 2013;**12**:319-360

[52] Chesson P. Mechanisms of maintenance of species diversity. Annual Review of Ecology and Systematics. 2000;**31**(1):343-366

[53] Tilman D. Resource Competition and Community Structure. Princeton University Press, Princeton, New Jersey, U.S.A. 1982

[54] Chase JM, Abrams PA, Grover JP, Diehl S, Chesson P, Holt RD, et al. The interaction between predation and competition: A review and synthesis. Ecology Letters. 2002; **5**(2):302-315

[55] O'Connor PJ, Smith SE, Smith FA. *Arbuscular mycorrhizas* influence plant diversity and community structure in a semiarid herbland. New Phytologist. 2002;**154**(1):209-218

[56] Reynolds HL, Packer A, Bever JD, Clay K. Grassroots ecology: Plant–microbe–soil interactions as drivers of plant community structure and dynamics. Ecology. 2003; **84**(9):2281-2291

[57] Simard SW, Durall DM. Mycorrhizal networks: A review of their extent, function, and importance. Canadian Journal of Botany. 2004;**82**(8):1140-1165

[58] Pietikäinen A, KYTÖVIITA MM. Defoliation changes mycorrhizal benefit and competitive interactions between seedlings and adult plants. Journal of Ecology. 2007;**95**(4):639-647

[59] Mitchell CE, Agrawal AA, Bever JD, Gilbert GS, Hufbauer RA, Klironomos JN, et al. Biotic interactions and plant invasions. Ecology Letters. 2006;**9**(6):726-740

[60] Stinson KA, Campbell SA, Powell JR, Wolfe BE, Callaway RM, Thelen GC, et al. Invasive plant suppresses the growth of native tree seedlings by disrupting belowground mutualisms. PLoS Biology. 2006;**4**(5):e140

[61] Vogelsang KM, Bever JD. Mycorrhizal densities decline in association with nonnative plants and contribute to plant invasion. Ecology. 2009;**90**(2):399-407

[62] Dickie IA, Schnitzer SA, Reich PB, Hobbie SE. Spatially disjunct effects of co-occurring competition and facilitation. Ecology Letters. 2005;**8**(11):1191-1200

[63] Reinhart KO, Callaway RM. Soil biota and invasive plants. New Phytologist. 2006;**170**(3):445-457

[64] Lundberg DS, Lebeis SL, Paredes SH, Yourstone S, Gehring J, Malfatti S, et al. Defining the core *Arabidopsis thaliana* root microbiome. Nature. 2012;**488**(7409):86-90

[65] Peiffer JA, Spor A, Koren O, Jin Z, Tringe SG, Dangl JL, et al. Diversity and heritability of the maize rhizosphere microbiome under field conditions. Proceedings of the National Academy of Sciences. 2013;**110**(16):6548-6553

[66] Bokulich NA, Thorngate JH, Richardson PM, Mills DA. Microbial biogeography of wine grapes is conditioned by cultivar, vintage, and climate. Proceedings of the National Academy of Sciences. 2014;**111**(1):E139-E148

[67] Bodenhausen N, Bortfeld-Miller M, Ackermann M, Vorholt JA. A synthetic community approach reveals plant genotypes affecting the phyllosphere microbiota. PLoS Genetics. 2014;**10**(4):e1004283

[68] Consortium HMP. Structure, function and diversity of the healthy human microbiome. Nature. 2012;**486**(7402):207-214

[69] Benson AK, Kelly SA, Legge R, Ma F, Low SJ, Kim J, et al. Individuality in gut microbiota composition is a complex polygenic trait shaped by multiple environmental and host genetic factors. Proceedings of the National Academy of Sciences. 2010;**107**(44):18933-18938

[70] Chaston JM, Dobson AJ, Newell PD, Douglas AE. Host genetic control of the microbiota mediates the Drosophila nutritional phenotype. Applied and Environmental Microbiology. 2016;**82**(2):671-679

[71] Hartmann A, Rothballer M, Schmid M. Lorenz Hiltner, a pioneer in rhizosphere microbial ecology and soil bacteriology research. Plant and Soil. 2008;**312**(1-2):7-14

[72] Lugtenberg B, Kamilova F. Plant-growth-promoting rhizobacteria. Annual Review of Microbiology. 2009;**63**:541-556

[73] Rovira AD. Plant root exudates. The Botanical Review. 1969;**35**(1):35-57

[74] Kirk JL, Klironomos JN, Lee H, Trevors JT. The effects of perennial ryegrass and alfalfa on microbial abundance and diversity in petroleum contaminated soil. Environmental Pollution. 2005;**133**(3):455-465

[75] İnceoğlu Ö, van Overbeek LS, Salles JF, van Elsas JD. Normal operating range of bacterial communities in soil used for potato cropping. Applied and Environmental Microbiology. 2013;**79**(4):1160-1170

[76] Li X, Zhang T, Wang X, Hua K, Zhao L, Han Z. The composition of root exudates from two different resistant peanut cultivars and their effects on the growth of soil-borne pathogen. International Journal of Biological Sciences. 2013;**9**(2):164-173

[77] Yuan J, Zhang N, Huang Q, Raza W, Li R, Vivanco JM, et al. Organic acids from root exudates of banana help root colonization of PGPR strain *Bacillus amyloliquefaciens* NJN-6. Scientific Reports. 2015;**5**:13438

[78] Bacilio-Jiménez M, Aguilar-Flores S, Ventura-Zapata E, Pérez-Campos E, Bouquelet S, Zenteno E. Chemical characterization of root exudates from rice (*Oryza sativa*) and their effects on the chemotactic response of endophytic bacteria. Plant and Soil. 2003;**249**(2):271-277

[79] Chen HY, Huh JH, Yu YC, Ho LH, Chen LQ, Tholl D, et al. The Arabidopsis vacuolar sugar transporter SWEET2 limits carbon sequestration from roots and restricts Pythium infection. The Plant Journal. 2015;**83**(6):1046-1058

[80] VanEtten HD, Mansfield JW, Bailey JA, Farmer EE. Two classes of plant antibiotics: Phytoalexins versus "Phytoanticipins". The Plant Cell. 1994;**6**(9):1191

[81] Vaughan MM, Wang Q, Webster FX, Kiemle D, Hong YJ, Tantillo DJ, et al. Formation of the unusual semivolatile diterpene rhizathalene by the Arabidopsis class I terpene synthase TPS08 in the root stele is involved in defense against belowground herbivory. The Plant Cell. 2013;**25**(3):1108-1125

[82] Lanoue A, Burlat V, Henkes GJ, Koch I, Schurr U, Röse US. De novo biosynthesis of defense root exudates in response to Fusarium attack in barley. New Phytologist. 2010;**185**(2):577-588

[83] Vuković R, Bauer N, Ćurković-Perica M. Genetic elicitation by inducible expression of β-cryptogein stimulates secretion of phenolics from Coleus blumei hairy roots. Plant Science. 2013;**199**:18-28

[84] Bais HP, Walker TS, Schweizer HP, Vivanco JM. Root specific elicitation and antimicrobial activity of rosmarinic acid in hairy root cultures of *Ocimum basilicum*. Plant Physiology and Biochemistry. 2002;**40**(11):983-995

[85] Toyomasu T, Kagahara T, Okada K, Koga J, Hasegawa M, Mitsuhashi W, et al. Diterpene phytoalexins are biosynthesized in and exuded from the roots of rice seedlings. Bioscience, Biotechnology, and Biochemistry. 2008;**72**(2):562-567

[86] Badri DV, Chaparro JM, Zhang R, Shen Q, Vivanco JM. Application of natural blends of phytochemicals derived from the root exudates of Arabidopsis to the soil reveal that phenolic-related compounds predominantly modulate the soil microbiome. Journal of Biological Chemistry. 2013;**288**(7):4502-4512

[87] Husain S, McKeen W. Interactions between strawberry roots and *Rhizoctonia fragariae*. Phytopathology. 1963;**53**(5):541-545

[88] Katznelson H, Rouatt J, Payne T. The liberation of amino acids and reducing compounds by plant roots. Plant and Soil. 1955;**7**(1):35-48

[89] Bowen G. Nutrient status effects on loss of amides and amino acids from pine roots. Plant and Soil. 1969;**30**(1):139-142

[90] Toljander JF, Santos-González JC, Tehler A, Finlay RD. Community analysis of arbuscular mycorrhizal fungi and bacteria in the maize mycorrhizosphere in a long-term fertilization trial. FEMS Microbiology Ecology. 2008;**65**(2):323-338

[91] Dumbrell AJ, Nelson M, Helgason T, Dytham C, Fitter AH. Relative roles of niche and neutral processes in structuring a soil microbial community. The ISME Journal. 2010;**4**(3):337-345

[92] Jin CW, Ye YQ, Zheng SJ. An underground tale: contribution of microbial activity to plant iron acquisition via ecological processes. Annals of Botany. 2014;**113**(1):7-18

[93] Rout ME, Southworth D. The root microbiome influences scales from molecules to ecosystems: The unseen majority. American Journal of Botany. 2013;**100**(9):1689-1691

[94] Harmsen G, Jager G. Determination of the Quantity of Carbon and Nitrogen in the Rhizosphere of Young Plants. In: Doeksen J, van der Drift J, editors. Soil Organisms. Amsterdam: North Holland; 1963;245-251

[95] Norman AG. Microbial products affecting root development. Transactions of the 7th Congress Wisconsin International Soil Science Society; 1960;**2**:531-536

[96] Pii Y, Mimmo T, Tomasi N, Terzano R, Cesco S, Crecchio C. Microbial interactions in the rhizosphere: Beneficial influences of plant growth-promoting rhizobacteria on nutrient acquisition process. A review. Biology and Fertility of Soils. 2015;**51**(4):403-415

[97] Vellend M, Agrawal A. Conceptual synthesis in community ecology. The Quarterly Review of Biology. 2010;**85**(2):183-206

[98] Hartmann M, Lee S, Hallam SJ, Mohn WW. Bacterial, archaeal and eukaryal community structures throughout soil horizons of harvested and naturally disturbed forest stands. Environmental Microbiology. 2009;**11**(12):3045-3062

[99] Smith P. Why Tiny Microbes Mean Big Things for Farming. National Geographic. 2014. Available from: http://news/ nationalgeographic com/news/2014/09/140918-soil-bacteria-microbe-farming-technology-ngfood

[100] Berendsen RL, Pieterse CM, Bakker PA. The rhizosphere microbiome and plant health. Trends in Plant Science. 2012;**17**(8):478-486

[101] Bulgarelli D, Rott M, Schlaeppi K, van Themaat EVL, Ahmadinejad N, Assenza F, et al. Revealing structure and assembly cues for Arabidopsis root-inhabiting bacterial microbiota. Nature. 2012;**488**(7409):91-95

[102] Agler MT, Ruhe J, Kroll S, Morhenn C, Kim S-T, Weigel D, et al. Microbial hub taxa link host and abiotic factors to plant microbiome variation. PLoS Biology. 2016;**14**(1):e1002352

[103] Foster R. Microenvironments of soil microorganisms. Biology and Fertility of Soils. 1988;**6**(3):189-203

[104] Lynch, J.M. The Rhizosphere. John Wiley & Sons, Chichester. 1990

[105] Glick BR. Plant growth-promoting bacteria: Mechanisms and applications. Scientifica. 2012;**2012**:1-15

[106] Grover M, Ali SZ, Sandhya V, Rasul A, Venkateswarlu B. Role of microorganisms in adaptation of agriculture crops to abiotic stresses. World Journal of Microbiology and Biotechnology. 2011;**27**(5):1231-1240

[107] Saeidnejad AH, Pouramir F, Naghizadeh M. Improving chilling tolerance of maize seedlings under cold conditions by spermine application. Notulae Scientia Biologicae. 2012;**4**(3):110

[108] Baek K-H, Skinner DZ. Production of reactive oxygen species by freezing stress and the protective roles of antioxidant enzymes in plants. Journal of Agricultural Chemistry and Environment. 2012;**1**:34-40

[109] Bradáčová K, Weber NF, Morad-Talab N, Asim M, Imran M, Weinmann M, et al. Micronutrients (Zn/Mn), seaweed extracts, and plant growth-promoting bacteria as cold-stress protectants in maize. Chemical and Biological Technologies in Agriculture. 2016;**3**(1):19

[110] Subramanian P, Kim K, Krishnamoorthy R, Mageswari A, Selvakumar G, Sa T. Cold stress tolerance in psychrotolerant soil bacteria and their conferred chilling resistance in tomato (Solanum lycopersicum Mill.) under low temperatures. PloS One. 2016;11(8):e0161592

[111] Verma P, Yadav AN, Khannam KS, Panjiar N, Kumar S, Saxena AK, et al. Assessment of genetic diversity and plant growth promoting attributes of psychrotolerant bacteria allied with wheat (*Triticum aestivum*) from the northern hills zone of India. Annals of Microbiology. 2015;**65**(4):1885-1899

[112] Barka EA, Nowak J, Clément C. Enhancement of chilling resistance of inoculated grapevine plantlets with a plant growth-promoting rhizobacterium, *Burkholderia phytofirmans* strain PsJN. Applied and Environmental Microbiology. 2006;**72**(11):7246-7252

[113] Ali SZ, Sandhya V, Grover M, Linga VR, Bandi V. Effect of inoculation with a thermotolerant plant growth promoting *Pseudomonas putida* strain AKMP7 on growth of wheat (*Triticum* spp.) under heat stress. Journal of Plant Interactions. 2011;**6**(4):239-246

[114] Ali SZ, Sandhya V, Grover M, Kishore N, Rao LV, Venkateswarlu B. Pseudomonas sp. strain AKM-P6 enhances tolerance of sorghum seedlings to elevated temperatures. Biology and Fertility of Soils. 2009;**46**(1):45-55

[115] Shahbaz M, Ashraf M. Improving salinity tolerance in cereals. Critical Reviews in Plant Sciences. 2013;**32**(4):237-249

[116] Shekar CC, Sammaiah D, Shasthree T, Reddy KJ. Effect of mercury on tomato growth and yield attributes. International Journal of Pharma and Bio Sciences. 2011;**2**(2):B358-B364

[117] Tank N, Saraf M. Salinity-resistant plant growth promoting rhizobacteria ameliorates sodium chloride stress on tomato plants. Journal of Plant Interactions. 2010;**5**(1):51-58

[118] Jha Y, Subramanian R, Patel S. Combination of endophytic and rhizospheric plant growth promoting rhizobacteria in *Oryza sativa* shows higher accumulation of osmoprotectant against saline stress. Acta Physiologiae Plantarum. 2011;**33**(3):797-802

[119] Bharti N, Pandey SS, Barnawal D, Patel VK, Kalra A. Plant growth promoting rhizobacteria *Dietzia natronolimnaea* modulates the expression of stress responsive genes providing protection of wheat from salinity stress. Scientific Reports 2016;**6**

[120] Qurashi AW, Sabri AN. Osmoadaptation and plant growth promotion by salt tolerant bacteria under salt stress. African Journal of Microbiology Research. 2011;**5**(21):3546-3554

[121] Mathew DC, Ho Y-N, Gicana RG, Mathew GM, Chien M-C, Huang C-C. A rhizosphere-associated symbiont, *Photobacterium* spp. strain MELD1, and its targeted synergistic activity for phytoprotection against mercury. PloS One. 2015;10(3):e0121178

[122] Farooq M, Wahid A, Kobayashi N, Fujita D, Basra S. Plant drought stress: effects, mechanisms and management. In: Sustainable Agriculture. Springer Netherlands. 2009;153-188

[123] Mayak S, Tirosh T, Glick BR. Plant growth-promoting bacteria that confer resistance to water stress in tomatoes and peppers. Plant Science. 2004;**166**(2):525-530

[124] Saravanakumar D, Kavino M, Raguchander T, Subbian P, Samiyappan R. Plant growth promoting bacteria enhance water stress resistance in green gram plants. Acta Physiologiae Plantarum. 2011;**33**(1):203-209

[125] Weast RC, Astle MJ, Beyer WH. CRC handbook of chemistry and Physics. CRC Press: Boca Raton, FL; 1988

[126] Lasat M. Phytoextraction of metals from contaminated soil: A review of plant/soil/ metal interaction and assessment of pertinent agronomic issues. Journal of Hazardous Substance Research. 2000;**2**(5):1-25

[127] Kamran MA, Syed JH, Eqani SAMAS, Munis MFH, Chaudhary HJ. Effect of plant growth-promoting rhizobacteria inoculation on cadmium (Cd) uptake by *Eruca sativa*. Environmental Science and Pollution Research. 2015;**22**(12):9275-9283

[128] Burd GI, Dixon DG, Glick BR. A plant growth-promoting bacterium that decreases nickel toxicity in seedlings. Applied and Environmental Microbiology. 1998;**64**(10):3663-3668

[129] Ho Y-N, Hsieh J-L, Huang C-C. Construction of a plant–microbe phytoremediation system: Combination of vetiver grass with a functional endophytic bacterium, *Achromobacter xylosoxidans* F3B, for aromatic pollutants removal. Bioresource Technology 2013;**145**:43-47

[130] Ho Y-N, Mathew DC, Hsiao S-C, Shih C-H, Chien M-F, Chiang H-M, et al. Selection and application of endophytic bacterium *Achromobacter xylosoxidans* strain F3B for improving phytoremediation of phenolic pollutants. Journal of Hazardous Materials. 2012;**219**:43-49

[131] Keel C, Schnider U, Maurhofer M, Voisard C, Laville J, Burger U, et al. Suppression of root diseases by *Pseudomonas fluorescens* CHA0: importance of the bacterial secondary metabolite 2, 4-diacetylphloroglucinol. Molecular Plant-Microbe Interactions.1992;**5**(1):4-13

[132] Lim H-S, Kim Y-S, Kim S-D. *Pseudomonas stutzeri* YPL-1 genetic transformation and antifungal mechanism against *Fusarium solani*, an agent of plant root rot. Applied and Environmental Microbiology. 1991;**57**(2):510-516

[133] Hill D, Stein J, Torkewitz N, Morse A, Howell C, Pachlatko J, et al. Cloning of genes involved in the synthesis of pyrrolnitrin from *Pseudomonas fluorescens* and role of pyrrolnitrin synthesis in biological control of plant disease. Applied and Environmental Microbiology. 1994;**60**(1):78-85

[134] Van Peer R, Niemann G, Schippers B. Induced resistance and phytoalexin accumulation in biological control of Fusarium wilt of carnation by Pseudomonas sp. strain WCS 417 r. Phytopathology. 1991;**81**(7):728-734

[135] Ardebili ZO, Ardebili NO, Hamdi SMM. Physiological effects of *Pseudomonas fluorescens* CHA0 on tomato (Lycopersicon esculentum Mill.) plants and its possible impact on *Fusarium oxysporum* f. sp. lycopersici. Australian Journal of Crop Science. 2011;**5**(12):1631

[136] Idriss EE, Makarewicz O, Farouk A, Rosner K, Greiner R, Bochow H, et al. Extracellular phytase activity of *Bacillus amyloliquefaciens* FZB45 contributes to its plant-growth-promoting effecta. Microbiology. 2002;**148**(7):2097-2109

[137] Kloepper JW, Ryu C-M. Bacterial endophytes as elicitors of induced systemic resistance. In: Microbial Root Endophytes. Springer Berlin Heidelberg, 2006;**9**:33-52

[138] Sen K, Sengupta C, Saha J. PGPR consortium alleviating downy mildew of cucumber. International journal of plant, animal and environmental sciences. 2014;**4**(4):150-159

[139] Lehr P. Biopesticides: The Global Market. Report No: CHM029C. Wellesley, Massachusetts: BCC Research; 2010

[140] Fawcett H. Fungus and bacterial diseases of insects as factors in biological control. The Botanical Review. 1944;**10**(6):327

[141] Roh JY, Choi JY, Li MS, Jin BR, Je YH. *Bacillus thuringiensis* as a specific, safe, and effective tool for insect pest control. Journal of Microbiology and Biotechnology. 2007;**17**(4):547

[142] Boets A, Arnaut G, Van Rie J, Damme N. Toxins. Google Patents; 2004

[143] De Oliveira EJ, Rabinovitch L, Monnerat RG, Passos LKJ, Zahner V. Molecular characterization of *Brevibacillus laterosporus* and its potential use in biological control. Applied and Environmental Microbiology. 2004;**70**(11):6657-6664

[144] Bone LW, Singer S. Control of parasitic nematode ova/larvae with a *Bacillus laterosporus*. Google Patents; 1991

[145] Singer S. The utility of strains of morphological group II Bacillus. Advances in Applied Microbiology. 1996;**42**:219-261

[146] Saikia R, Gogoi D, Mazumder S, Yadav A, Sarma R, Bora T, et al. Brevibacillus laterosporus strain BPM3, a potential biocontrol agent isolated from a natural hot water spring of Assam, India. Microbiological Research 2011;**166**(3):216-225

[147] Smith VH, Tilman GD, Nekola JC. Eutrophication: Impacts of excess nutrient inputs on freshwater, marine, and terrestrial ecosystems. Environmental Pollution. 1999;**100**(1):179-196

[148] Vorholt JA. Microbial life in the phyllosphere. Nature Reviews Microbiology. 2012; **10**(12):828-840

[149] Bulgarelli D, Schlaeppi K, Spaepen S, van Themaat EVL, Schulze-Lefert P. Structure and functions of the bacterial microbiota of plants. Annual Review of Plant Biology. 2013;**64**:807-838

[150] Reid A. Microbes helping to improve crop productivity. Microbe. 2011;**6**(10):435

[151] Denison RF, Kiers ET. Life histories of symbiotic rhizobia and mycorrhizal fungi. Current Biology. 2011;**21**(18):R775-R785

[152] Vejan P, Abdullah R, Khadiran T, Ismail S, Nasrulhaq Boyce A. Role of plant growth promoting rhizobacteria in agricultural sustainability—A review. Molecules. 2016;**21**(5):573

[153] de Souza EM, Granada CE, Sperotto RA. Plant pathogens affecting the establishment of plant-symbiont interaction. Frontiers in Plant Science 2016;**7**

[154] Soda N, Wallace S, Karan R. Omics study for abiotic stress responses in plants. Advances in Plants & Agriculture Research. 2015;**2**(1):00037

[155] Gupta B, Saha J, Sengupta A, Gupta K. Plant Abiotic Stress: 'Omics' Approach. Journal of Plant Biochemistry and Biotechnology. 2013;**1**:e108

[156] Barkal LJ, Theberge AB, Guo C-J, Spraker J, Rappert L, Berthier J, et al. Microbial metabolomics in open microscale platforms. Nature Communications 2016;**7**:10610

[157] Narasimhan K, Basheer C, Bajic VB, Swarup S. Enhancement of plant-microbe interactions using a rhizosphere metabolomics-driven approach and its application in the removal of polychlorinated biphenyls. Plant Physiology. 2003;**132**(1):146-153

[158] Thynne E, McDonald MC, Solomon PS. Phytopathogen emergence in the genomics era. Trends in Plant Science. 2015;**20**(4):246-255

[159] Schneider DJ, Collmer A. Studying plant-pathogen interactions in the genomics era: Beyond molecular Koch's postulates to systems biology. Annual Review of Phytopathology. 2010;**48**:457-479

[160] Salanoubat M, Genin S, Artiguenave F, Gouzy J, Mangenot S, Arlat M, et al. Genome sequence of the plant pathogen *Ralstonia solanacearum*. Nature. 2002;**415**(6871):497-502

[161] Simpson AJG, Reinach F, Arruda P, Abreu F, Acencio M, Alvarenga R, et al. The genome sequence of the plant pathogen *Xylella fastidiosa*. Nature. 2000;**406**(6792):151-157

[162] Durot M, Bourguignon P-Y, Schachter V. Genome-scale models of bacterial metabolism: Reconstruction and applications. FEMS Microbiology Reviews. 2009;**33**(1):164-190

[163] Magnúsdóttir S, Heinken A, Kutt L, Ravcheev DA, Bauer E, Noronha A, et al. Generation of genome-scale metabolic reconstructions for 773 members of the human gut microbiota. Nature Biotechnology. 2017;**35**:81-89

[164] Nogales J, Palsson BØ, Thiele I. A genome-scale metabolic reconstruction of *Pseudomonas putida* KT2440: iJN746 as a cell factory. BMC Systems Biology. 2008;**2**(1):79

[165] Timm CM, Campbell AG, Utturkar SM, Jun S-R, Parales RE, Tan WA, et al. Metabolic functions of *Pseudomonas fluorescens* strains from *Populus deltoides* depend on rhizosphere or endosphere isolation compartment. Frontiers in Microbiology 2015;**6**:1118

[166] Wu CH, Bernard SM, Andersen GL, Chen W. Developing microbe–plant interactions for applications in plant-growth promotion and disease control, production of useful compounds, remediation and carbon sequestration. Microbial Biotechnology. 2009;**2**(4):428-440

[167] Kroll S, Agler MT, Kemen E. Genomic dissection of host–microbe and microbe–microbe interactions for advanced plant breeding. Current Opinion in Plant Biology. 2017;**36**:71-78

[168] Bender CL. The post-genomic era: New approaches for studying bacterial diseases of plants. Australasian Plant Pathology. 2005;**34**(4):471-474

[169] Quirino B, Candido E, Campos P, Franco O, Krüger R. Proteomic approaches to study plant–pathogen interactions. Phytochemistry. 2010;**71**(4):351-362

[170] Winnenburg R, Urban M, Beacham A, Baldwin TK, Holland S, Lindeberg M, et al. PHI-base update: additions to the pathogen–host interaction database. Nucleic Acids Research. 2008;**36**(Suppl 1):D572–D576

Chapter 7

Phytosociological Surveys in Weed Science: Old Concept, New Approach

Germani Concenço, Patrícia Menegaz de Farias,
Néstor Fabio Alzate Quintero, Fábio Schreiber,
Leandro Galon, Michely Tomazi,
Ivana Santos Moisinho,
Mariane Camponogara Coradini,
William Christofari Ceolin and André Andres

Additional information is available at the end of the chapter

http://dx.doi.org/10.5772/intechopen.69083

Abstract

Phytosociological surveys have been applied to studies on agroecosystems, especially in relation to weed populations into arable fields. These surveys can indicate trends of variation of the importance of plant populations within a crop, and whether the variations are associated to agricultural practices adopted, which can be further used to support the development of weed management programs. However, to understand the applicability of phytosociological studies for weeds, it is necessary to understand the ecological basis and determine the most appropriate methods to be used when surveying arable fields. Therefore, the aim of the present chapter is to introduce a new approach of phytosociological survey to be used as a tool for the weed science. Throughout the chapter, this new approach is presented in details covering aspects related to methods for sampling and describing weed communities. The following sequence of steps is proposed as the most suitable for a weed phytosociological and association survey: (1) overall infestation; (2) phytosociological tables/graphs; (3) intra-characterization by diversity; (4) inter-characterization and grouping by multivariate analysis; and (5) weeds association through contingency tables.

Keywords: weed species, survey, data processing, diversity, sustainability, agriculture

1. Introduction

The classification of plant species is necessary to understand the complexity of environments, being based mainly on morphology and recently aided by genetics and its functional properties [1]. Plant communities are a set of plant species within a given geographic unit, which form relatively uniform patches, distinguishable from patches of different types of vegetation adjacent to that limited area [2]. Vegetation ecology seeks to identify the species found within the same habitat, thus, describing the physiognomy of the landscape, in order to determine why the communities have a given structure as well as their mechanisms of adaptation [3]. When it comes to community studies, it is necessary to understand how environmental conditions and species interactions influence the patterns of coexistence and relative abundance of species on the local scale, but it should be taken into account the important role of spatio-temporal dynamics [4–6].

The classification of plant communities into a hierarchical system is made in an inductively synthetic way; the types of vegetation units are abstracted as basic syntaxonomic units and compiled as associations [2, 7]. The basis of the phytosociological categorization of plants, according to Blasi and Frondoni [8], is drawn into a context of botanical geography, with the primordial observation that plant species are grouped in associations in which they differ in composition and/or physiognomy, according to geographic regions and environmental conditions (e.g., climate, altitude, latitude).

In this sense, phytosociology is the science that seeks to understand, through the composition and distribution of plant species in a given phytogeographic region, the diversity of plant communities [9]. However, this is a phytogeographic and non-phytosociological approach [10], since the phytosociological approach in the original Braun-Blanquet concept is a floristic statistic based on the occurrence of plant species. In the concept proposed by Oosting [11] and Harper [12], phytosociology is the science of plant communities as well as the relationships with the environment and the processes that modify these communities. This approach is more related to functional ecology, which seeks to understand how and why ecological systems, and their components interact differently in different environments [13].

The term phytosociology, however, is directly associated with the structure of a community of plant species [14], while phytocenosis is defined as the study of plant cover [11]. In this sense, when it comes to a phytosociological survey through an inductive statistical process of inventory comparison, it is necessary to establish a conceptual class that represents a model of phytocenosis. Another concept refers to the possibility of representing patterns in floristic structure and combination, thus modeling phytocenosis as a resource for the dynamic systems theory [15, 16].

A phytosociological study considers three principles: the **analytical** (portrays the size of the inventory surface, characteristics of the sampling site, and variables such as abundance, density, dominance, and the sociability of plant species), the **synthetic** (referring to frequency of species which compose the plant community), and **syntaxonomy** (establishing the phytosociological hierarchy) [9, 11, 15, 16]. Thus, for a phytosociological study, it is necessary to take into account that vegetation varies in spatial scales, which are defined by the size of the sampled units and that several patterns can be detected only with the change in the observation scale, so the size of the sampling unit influences the analysis of spatial patterns of plant populations [17]. In this way, evaluating the quality and quantity in the composition of species of phytocenosis, as an expression of all historical, sociological, and local influences of the abiotic factors, can be a key point for understanding the floristic composition [2].

Basically, the procedures and methods for sampling and registering plant communities follow the Braun-Blanquet [9] method as it considers the analysis and description of selected plant populations as basic types [18]. This method, however, has some limitations and alternatives have been developed along the years [14].

Phytosociological surveys have been applied to studies on agroecosystems, especially in relation to weed plant populations into crops [19]. An infesting population is a result of the interactional relationship between phenotypic plasticity of each individual and long-term processes that provided adaptive flexibility to eventual changes in the natural or artificial environment [20, 21]. Thus, when conducting a phytosociological study of weed species in crops, these can indicate trends of variation of the importance of plant populations within a crop, and these variations may be associated to the agricultural practices adopted as well as subsidize the development of weed management programs.

In general, the phytosociological studies of weed plant communities in agroecosystems allow the determination of periods of control and/or coexistence between crop and weeds, and through the phytosociological indices, it is possible to determine which species are the most important in the different periods of growth of the weed community [14]. However, to understand the applicability of phytosociological studies for weed species in crops, it is necessary to understand the ecological basis and determine the most appropriate methods to be used when surveying arable fields.

1.1. Two contrasting theories

The study of vegetation played an important role in the evolution of ecological concepts through the formulation of several vegetation theories as well as methods of surveying and analyzing phytosociological data [10, 22]. In phytosociological terms, the concept of community is based on the principle of associations (different groupings of plant species, usually found in sites with similar environmental conditions) [14].

The definition of associations proposed by some authors [1, 23] considers the type of vegetation that represents the real plant communities and shares a certain combination of statistically reliable characteristics, in terms of physiognomy and stratification, ecological conditions, dynamic meaning, area of distribution, and history. This gives the association a greater value of information in ecological and geographical terms, which increases the indicator value of vegetation in a site [8]. However, there is some discussion in terms of synecological methods related to the concept of community. For instance, the concept of community proposed by Begon [24] refers to a set of species that inhabit the same area at a given time, while Gurevitch [25] refers to a group of populations that coexist in space and time, interacting directly or indirectly.

Two great ecologists, Frederick E. Clements and H.A. Gleason, presented a series of discussions on community ecology. The driving issue was whether the community was a self-organized system of co-occurring species or simply a random collection of populations with minimal functional integration. Two extreme views prevailed: one view considered a community as a "super-organism" whose species were strongly united by interactions that contributed to repetitive patterns of species abundance in space and time; in contrast, communities were a result of interactions among species, as well as between species and environment, combined with historically extreme and occasional climatic events [26].

For Clements, in his theory called "super-organisms", organisms and communities not only have their own growth and development but also evolve from predecessor communities. They supposedly have an ontogeny that one could study, just as is done with individuals and species, so one could classify the communities in a way comparable to the Linnaean taxonomy. It ultimately assumed a common evolutionary history for the integrated species [27], and the emergence and disappearance of a particular plant community was supposed to be easily and accurately estimated because it was considered as a single organism [26]. In short, the concept of plant community in this theory is defined as an autonomous, discrete, individualizable entity possessing its own structural and functional properties [27, 28]. This theory predicted that the optimum and the amplitude of the species presented distinct clusters, so expected changes in the vegetation would be abrupt [14].

On the other side, the theory of Henry A. Gleason focused on the traits of individual species that allow each to be within specific habitats or geographic areas [29]. This is a much more arbitrary unity than that imagined by Clements, since in Gleason's view, the spatial barriers of communities are not clear and assemblies of species can change considerably over time and space allowing each species to have its own tolerance to certain selection factors, thus, responding to environmental stresses in particular ways [26]. Thus, it was proposed a concept of a plant continuum in which species combinations result mostly from individual responses to environmental factors, and the occurrence of dispersion of individuals is random as a response to environmental fluctuations [28]. This theory states that the level of occurrence of a given plant species is proportional to the level of the stress the species can tolerate [25].

The theoretical opposition allowed the origin of two strands of current vegetation studies, the *discrete* versus *continuum* [26]. The same authors point out that the widely accepted view of the nature of communities is much closer to Gleason's view, since a given species may occur in species assemblies or communities, under different circumstances. When analyzing the relative importance that each of these authors gives to the different orders of factors in the vegetation theories, it is concluded that all of them consider the responses of the species to the habitat as the dominant influence in the structuring of the vegetation [28]. Begon [24] found that community ecology can be described as the study of the level of community organization rather than a spatially and/or temporally definable unit. Thus, it can be defined that the gradient of plant composition of the agglomerates is defined by the environment (or management in arable areas), and abrupt changes are observed in the composition of species within clusters when abrupt selection factors are applied [25, 26, 30, 31].

2. Aims and methods

Before reviewing the main methods and criteria involved in the sampling and evaluation of flora and vegetation, it is necessary to define some basic concepts:

1. Flora is the set of plant species present in a given place or area;
2. Vegetation refers to the quantitative aspects of plant architecture, that is, its horizontal and vertical distribution on the surface; and
3. The plant community should be understood as a set of plants of two or more plant species that coexist in a certain area, and according to the dominance of some of its species, it can be differentiated from other natural and/or altered plant communities [32, 33].

Based on Whittaker [34], "*Within phytosociological studies, the species inventory is fundamental to characterize both α diversity (species richness of a particular community considered homogeneous) and β diversity (degree of variability or replacement in the composition of species among different communities of an environment)*". Thus, in initiating a study of the composition, structure, and ordering of a plant community, the fundamentals are the choice of sampling type (random or systematic), location, size, shape, and quantity of sample units [26]. Once these are obtained, the researcher can apply the measurement of the attributes of the vegetation to be characterized (such as abundance, frequency, density, and dominance), biotypes, vertical structure, horizontal structure (coverage), and others [25, 26, 30]. The measurement of these attributes allows obtaining the importance value (IV) and biodiversity indices [26].

The purpose of phytosociological studies in the weed science does not differ much from the ecological field and rather tends to combine efforts between two disciplines (botany and ecology), in order to improve agricultural productivity and decrease the competition between crops and weeds [35]. Many of the traditional studies in weed science carried out in nonindustrialized (usually considered also under developed) countries have focused on adopting foreign technologies, with little research on biological and ecological aspects of weeds, diagnosis of population dynamics, and integrated weed management [36]. In this context, the use of phytosociological methods in the weed science can be directly associated with the nature of the treatments applied to arable fields, its intrinsic factors, and the history of the area where it will be established [14]. In arable fields, however, two implications must be considered:

1. The plots are usually much smaller in size than expected in phytosociological sampling of wild ecosystems and
2. There is a stronger set of factors influencing arable fields, such as plant density and height, history of land use, soil tillage, and application of agrochemicals, compared to natural environments.

Finally, communities of invasive plants show a behavior similar to that proposed by Gleason, in his theory of the individualistic concept of vegetable association, which states that "*Communities are the result of the interaction between individual species and their (biotic and abiotic) environment in combination with historical events*" [29].

2.1. Methods for sampling the community

The choice between the different variants of the methodologies depend on the sampling objective and the characteristics of the populations (richness and distribution) to be evaluated in each particular agroecosystem [9, 26]. Conventionally, the weed sampling methodologies do not always have prior information about these characteristics or some type of support in the decision-making process or sometimes they are not convenient. For example, on the number of points to be taken, conventional weed population sampling methodologies assume that they have homogeneous or random distribution in space [26] and this is not always the case [37] as numerous studies claim that weed distribution is in patches [38, 39]. It is necessary to emphasize that the sampling of weeds has two objectives:

1. Knowledge on the community richness and abundance, which in turn provides information for biodiversity studies (richness and structure of the communities) on the medium and long-term weed-management plans and

2. Mapping and spatial dynamics studies [9].

Some authors [26, 30] point out several sampling methods, but taking into account the limitations imposed by sampling arable fields, only two of them will be dealt with in this chapter: relevé and random quadrats.

2.1.1. Relevé

Braun-Blanquet [9] made an analogy between organisms and communities by comparing a species with a plant community for the purpose of establishing a classification of communities similar to the way organisms are classified into taxonomic groups. For him, the plant community is the basic unit of the taxonomic classification, which serves to establish a hierarchical system of classification of the communities on a world scale. The same author proposed that the selection of the area to be sampled should be carried out through the determination of the *minimum area*, which is defined as the smallest area where the floristic composition of the community is adequately represented. In the case of the relevé method, where the emphasis is given by the list of species, the minimum area is an indicator of the area needed to have a good sample of the community. The minimum area depends primarily on floristic diversity, plant size, and spacing among them in each community. The same is calculated in the field as follows:

1. A small area, say 0.25 m^2, is delineated, and the list of species present on that surface is recorded;

2. By adding the same area to the original one (2× size), the number of species should be recorded again, and the total number of species in the new quadrat should be counted again;

3. This should be done repeatedly (increase the area, count the species) until the number of species tends to stabilization; and

4. The values of the cumulative total of species (on the ordinate) corresponding to each of the successively duplicated areas (on the abscissa) are represented on a pair of perpendicular axes (**Figure 1**).

A strong slope in its initial part normally characterizes the resulting curve because the first areas incorporate a larger number of new species. Subsequently, as the sampled surface is increased, the appearance of new species in the quadrat becomes rarer and, consequently, the slope of the curve decreases tending to stabilization (**Figure 1**). The appropriate size of the sample unit should be found in the horizontal portion of the curve, and the point of inflection of the sample unit (when it is manifest) projected on the axis of the abscissa will indicate the minimum area. In general, it is convenient to use a size that exceeds a little the minimum area.

It is proposed also that each species in the list is accompanied by an estimate of its abundance-dominance by using the combined coverage-abundance scale and also by its degree of sociability (**Table 1**), both stated by Braun-Blanquet [9].

2.1.2. Random quadrats

In certain communities, the determination of frequency estimators (abundance, coverage) depends too much on the criteria of the expert in charge of the evaluation [26, 40], especially in herbaceous formations such as meadows, pastures, or high wetlands, in which it is most useful to use the method known as "random quadrat". This consists of finding subjective patterns within the community to be sampled and to conduct sampling in such a way so as not to favor

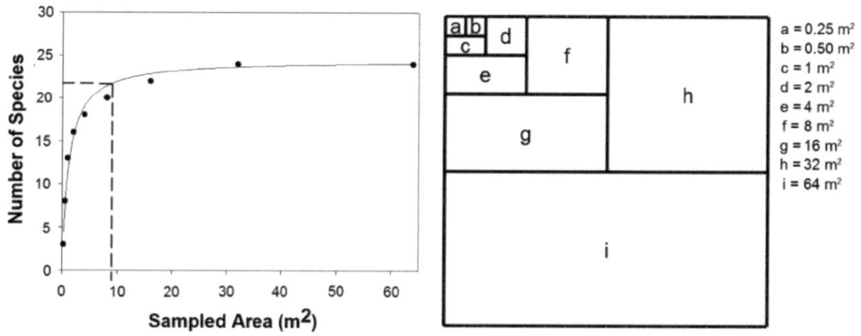

Figure 1. Calibration of the relevé method—determination of the minimal area of the single quadrat to be sampled for fidelity in terms of the number of weed species. Source: adapted from Concenço [14].

Score	Coverage-abundance		Sociability
	Number of individuals	Area coverage	
5	Any number	>75%	Large, almost pure stands
4	Any number	50–74%	Small colonies or carpets
3	Any number	25–49%	Small patches or cushions
2	Any number	5–24%	Small but dense clumps
1	Numerous	<5%	Growing singly
+	Few	Small	–
r	Scattered individuals	Small	–

Source: Adapted from Barbour [26] and Moore [40].

Table 1. Coverage abundance and sociability scales of Braun-Blanquet.

a particular pattern [26, 30]. It means that for the data to be reliable, sampling should be performed as randomly as possible. Several methods are available to help the researcher to go through and sample the area properly, but three of these methods are highlighted to be used in weed science: even spaced, by chance, and random by zones (**Figure 2**).

The geometric forms of the sample unit (called "quadrat") are basically of three types: square, rectangular, or circular [26]. These types of units allow the registration of all variables of dominance, frequency, and density of plant individuals [30]. Based on Goodall [41], surface geometry affects two aspects that significantly influence the vegetation sampling result.

First, let us consider the magnitude of the edge effect given by the area/perimeter ratio of the sample [26]. In the case of plots with rectangular shapes, the long-wide relation of the units and their directional position/orientation influence the degree of heterogeneity registered into the plot. To be considered part of the unit, the plants must be rooted within the perimeter, and the perimeter is longer in a rectangular quadrat compared to square or circular forms. Longer perimeters may increase the chance of the observer to be mistaken when deciding if a plant in the very border of the quadrat is actually in or out of the quadrat [26, 40]; this type of error is called "edge effect" [26].

If rooting occurs outside the plot area but its shoots occupy the airspace of the unit, the plant can be optionally registered as present, depending on the purpose of the survey [25, 32]. It is necessary, however, to make explicit the criterion established when defining the variable, but usually only plants rooted into the quadrat are considered.

2.2. Methods for describing the community

There is a wide variety of methods that allow the floristic characterization of a plant community, whose suitability or applicability depends on the specific objectives of each study and the

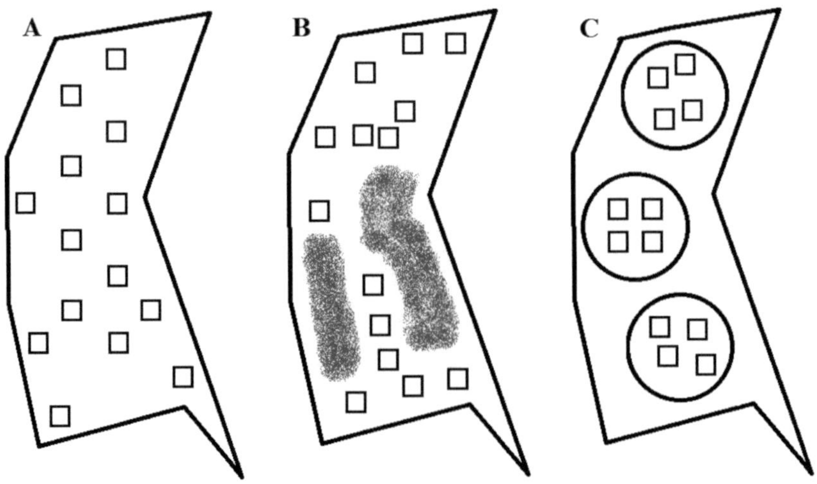

Figure 2. Distribution of samples for the random quadrats method. (A) Even spaced distribution, (B) chance distribution, and (C) random distribution of quadrats by zones. Source: adapted from Concenço [14].

structure of the community studied. However, and regardless of the method used for the floristic study, each sampling unit (quadrat) must meet the following criteria [42]:

1. It must be of sufficient size to contain the most possible proportion of species belonging to the plant community;
2. The habitat must be uniform into the sampling area, within the levels one can determine; and
3. Plant cover should be as homogeneous as possible.

2.2.1. *Importance components*

A fundamental aspect in the floristic characterization of a plant community is that the methodology adopted should provide an adequate representation of all the species present in the community in natural ecosystems [26]. For arable fields, one may hypothesize that it should properly represent at least most of the weed species present. Once field sampling is accomplished and all data are collected, the following parameters can be calculated [14]:

$$rDe = \frac{I}{TI} \times 100 \tag{1}$$

$$rFr = \frac{Q}{TQ} \times 100 \tag{2}$$

$$rDo = \frac{DM}{TDM} \times 100 \tag{3}$$

$$IV = \frac{rDe + rFr + rDo}{3} \tag{4}$$

where rDe = relative density (%); rFr = relative frequency (%); rDo = relative dominance (%); IV = importance value (%); I = number of individuals of species x in area r; TI = total number of individuals in area r; Q = number of samples evaluated in area r where species x is present; TQ = total number of samples in area r; DM = dry mass of individuals of species x in area r; and TDM = total dry mass of weeds in area r.

The *IV* locates each weed species within the community, depending on its ability to cause damage (severity of occurrence) based on the three parameters previously mentioned. In **Figure 3**, the nature of the importance components is illustrated.

2.2.2. *Diversity indices*

The calculation of diversity indices, α (alpha), β (beta), and γ (gamma), allows the comparative analysis of homogeneous or heterogeneous plant formations. They measure, respectively, the species richness of a community, the degree of change or replacement in species composition among different communities, and their richness in the set of communities [34, 40].

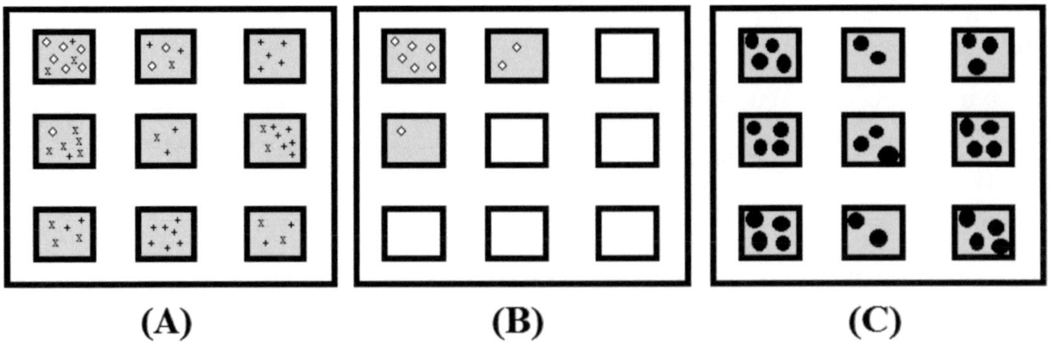

Figure 3. (A) Density, related to the number of plants of a given species found in all quadrats (each symbol is a different weed species); (B) frequency, related to the number of quadrats where a given species was found, independently of number of individuals; and (C) dominance, related to the amount of space in the canopy attributed to a given species, in arable fields, measured usually by dry mass accumulation. Source: adapted from Concenço [14].

The most widely used diversity indices are Margalef (α), Menhinick (*Dm*), Simpson (*D*), and modified Shannon–Weiner (*H*'), besides species density itself [25]. The Margalef index [43] is used to estimate the biodiversity of a community based on the numerical distribution of the individuals of the different species according to the number of individuals in the sample analyzed. It is obtained by Eq. (5):

$$\alpha = \frac{(s-1)}{\ln N} \tag{5}$$

where α = Margalef index, S = number of species, and N = total number of individuals.

This method can determine the number of taxa and the number of individuals in an ecosystem, comparing species richness among samples collected from different habitats.

The Menhinick index (*Dm*) is based on the relationship between the number of species and the total number of individuals observed, which increases together with sample size [34]. It is obtained by Eq. (6):

$$Dm = \frac{S}{\sqrt{N}} \tag{6}$$

where Dm = Menhinick index, S = species collected, and N = total number of individuals as sum of all species "*S*".

The Simpson index [26] is obtained by Eq. (7). Its calculation is strongly influenced by the importance of the most dominant species. Since its value is inverse to equity, diversity by Simpson is usually calculated by Eq. (8), which indicates that closer to the value of "1", the greater the equity. Simpson's *D* gives very little weight to rare species and is more sensitive to abundant species (those with greater number of individuals). The equations are:

$$\lambda = \sum Pi^2 \tag{7}$$

$$D = 1 - \sum Pi^2 \tag{8}$$

where λ = Simpson index, Pi = proportion of individuals of species "i" divided by the total number of individuals in the sample, and D = diversity of Simpson.

The Shannon–Weiner [26] diversity index (H) is another index commonly used to characterize species diversity in a community and is more sensitive to rare species; this is where sampling errors may be greater [26, 30, 40]. It is calculated by Eq. (9):

$$H' = -\sum [Pi \times \ln(Pi)] \tag{9}$$

where H' = Shannon–Weiner diversity index, Pi = proportion of individuals of species "i" divided by the total number of individuals in the sample.

It is a relationship between abundance and richness and expresses the uniformity of abundance values across all species in the sample. It ranges from "0", when there is only one species and the Neperian logarithm of "S" (number of species collected), when all species are represented by the same number of individuals [40].

In addition to the diversity indices, the Shannon–Weiner evenness proportion (SEP) sustainability coefficient [44], Eq. (11), is able to infer about sustainability of managements applied to production systems from static data. It considers the diversity of Shannon–Weiner calculated both from density (Eq. (9)) and from dry mass data, Eq. (10). As it is a division of one by the other, differences between H' and Hm' near zero, which correspond to SEP near "1", indicate longevity of the management practice applied and consequently of the production system, by the absence of strong species selecting factors. A visual representation of D, H', and SEP is supplied in **Figure 4**.

$$Hm' = -\sum [Mi \times \ln(Mi)] \tag{10}$$

$$SEP = \frac{Hm'}{H'} \tag{11}$$

where Hm' = Shannon–Weiner diversity index based on dry mass and Mi = dry mass of individuals of species "i" divided by the total dry mass of individuals in the sample.

2.2.3. *Multivariate analysis*

Descriptive multivariate analysis provides complementary tools to phytosociology. In this sense, classification and ordering techniques allow the identification of variation patterns in large data sets using algebraic procedures that can be translated into mathematical algorithms. Consequently, these techniques facilitate the work of comparing large sets of data from surveys with the help of computer programs.

Samples of plant communities, whether they are described by the presence or by abundance of the species that compose them, are multivariate because they present values of different variables (species) in each of the studied sites [46, 47]. Based on Matteucci and Colma [48] and Moreno [49], the degree of species turnover (beta diversity) has been evaluated mainly considering proportions

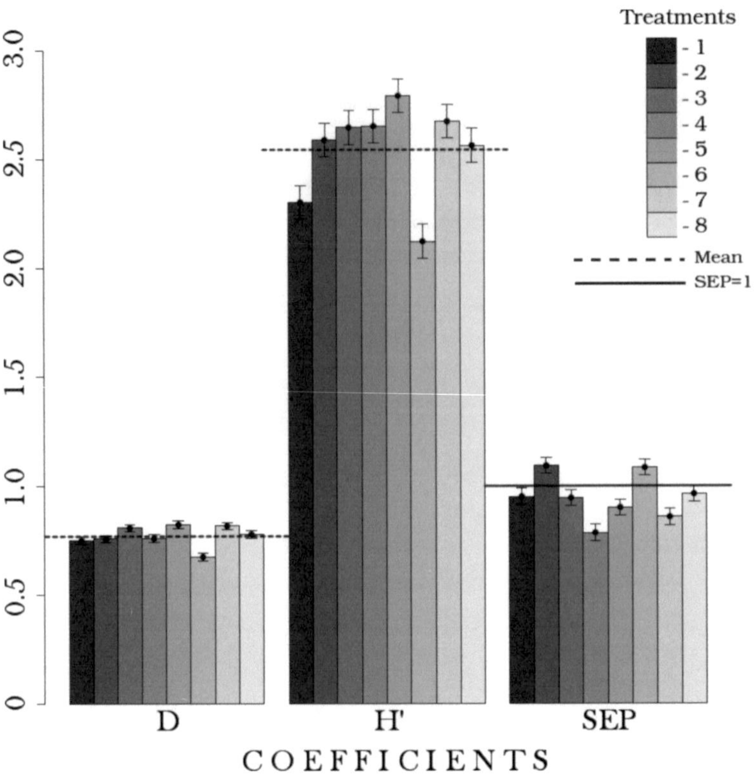

Figure 4. Application of the diversity coefficients of Simpson (*D*) and Shannon–Weiner (*H'*), as well as the sustainability coefficient *SEP*, to eight hypothetic treatments. Error bars are presented. Source: adapted from Concenço [45].

or differences. The proportions can be evaluated using indices as well as coefficients that indicate how similar/dissimilar two communities or samples are. Many of these similarities and differences can also be expressed or visualized by distances. These similarities or differences can be either qualitative (using presence-absence data) or quantitative (using proportional abundance data for each species or study group, as number of individuals, biomass, relative density, coverage, etc.).

The methods for quantifying beta diversity can be divided into two classes: similarity-dissimilarity and exchange/replacement of species. The different indices considered in the methods should be applied depending on the nature of the data (qualitative/quantitative) and what the relationship between the samples is, what it implies, how samples are organized, and how they were obtained, according to the question of interest. Thus, the similarity or dissimilarity expresses the degree of comparability in species composition and its abundances between two samples (communities).

2.2.4. Clustering by similarity

The beta diversity indices of Jaccard (*J*) and Sørensen (*So*) facilitate the comparison of areas in the composition of weed communities [50]. According to Concenço [14], these indices are considered high when they are above 0.25 (25%) and 0.5 (50%), respectively, in which a high resemblance between the areas can be interpreted. Booth [37] indicates that values should be interpreted on an absolute scale from 0 to 1, where 0 indicates total dissimilarity and 1 indicates absolute similarity.

Sørensen's index, Eq. (13), puts more weight on the co-occurrence of species, compared to Jaccard's index (Eq. (12)). Sørensen relates the number of shared species with the arithmetic mean of the species in both compared sites, while Jaccard relates the number of shared species with the total number of exclusive species [26, 40].

$$J = \frac{c}{(a + b - c)} \tag{12}$$

$$So = \frac{2c}{(a + b)} \tag{13}$$

where J = Jaccard's similarity index; So = Sørensen's similarity index; a = total number of weed species in area "a"; b = total number of weed species in area "b"; and c = number of weed species common to areas "a" and "b".

To group the areas according to their similarity, it is advised to first obtain the dissimilarity (differences) between areas (Di), by obtaining the distance from the values of J or So and "1" (Eqs. (14) and (15)):

$$Di_J = 1 - J \tag{14}$$

$$Di_{So} = 1 - So \tag{15}$$

where Di = dissimilarity (by Jaccard or Sørensen); J = Jaccard's similarity index; and So = Sørensen's similarity index.

After obtaining the dissimilarity matrix of treatment versus treatment (**Table 2**), multivariate analysis of hierarchical clustering may be performed by the unweighted pair group method with arithmetic mean (UPGMA) hierarchical clustering method [51] (**Figure 5**). The critical level for separation of groups in the cluster analysis is advised to be based on the arithmetic mean of the original matrix [26], disregarding crossing points between the same areas. Group validation is

Treat.	T1	T2	T3	T4
T1	0	0.64	0.59	0.53
T2	0.64	0	0.58	0.50
T3	0.59	0.58	0	0.46
T4	0.53	0.50	0.46	0

Table 2. Hypothetical dissimilarity matrix based on Jaccard's or Sørensen's similarity coefficients for four hypothetical treatments. Source: adapted from Concenço [14].

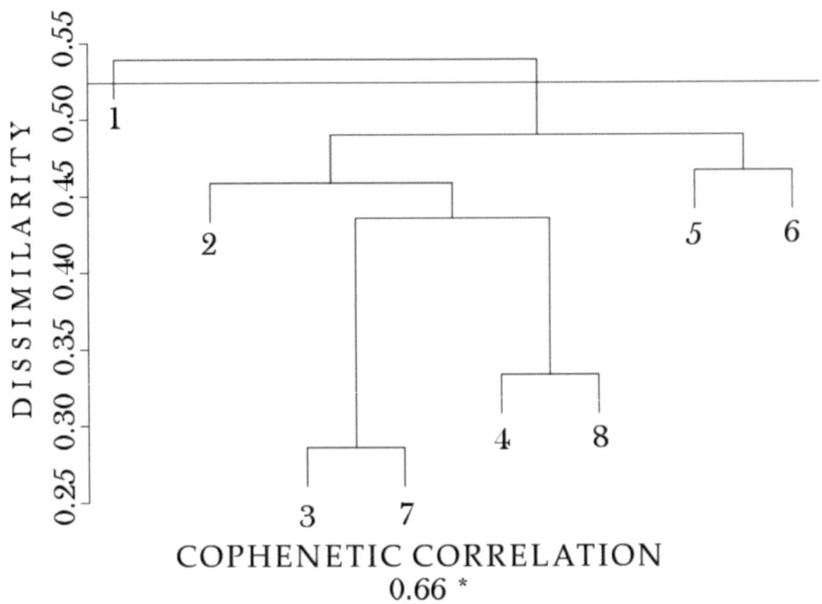

Figure 5. Cluster analysis for eight hypothetical treatments based on the dissimilarity matrix by Jaccard's coefficient and grouping accomplished by the UPGMA method. Source: adapted from Concenço [45].

usually accomplished by the cophenetic correlation coefficient [52], obtained by Pearson's linear correlation between the original matrix of dissimilarity and its respective cophenetic matrix.

3. Association between plant species

The relationships between weed species and the crop in arable fields are described by several authors in terms of competitive aspects [53, 54] and crop-yield losses [55–57]. The balancing in occurrence of weed species in these same arable fields is usually described by means of phytosociological surveys, as previously reported [14, 19, 38], being used as auxiliary to the competitive data aiming to subsidize recommendations for weed control in arable fields. But beyond a simple characterization of both crop losses by competition and the composition of weed occurrence, there is need to understand how weeds interact among them [25].

The principle of interaction among plant species is based on *associations*, which are different clusters of plant species, found generally together, in sites with similar conditions [26]. The first theory regarding plant association was proposed by Clements [27], which stated that plant communities are very organized entities. Thus, the emergence and disappearance of a given plant community could be precisely estimated [26, 27]. In contrast to Clements's ideas, Gleason [25, 29] reported that each species had its own tolerance to given selection factors; thus, they answered to environmental stresses in particular ways. Gleason, however, did not negate at all the occurrence of plant associations, defending that these associations would be more linked to environmental stresses and resource availability than to intrinsic plant traits [29, 40].

Currently, it is believed that plant association exists to a certain degree; the gradient of plant composition of weed clusters is defined by the environment (and by management in arable areas) and abrupt changes in plant composition into clusters are observed when abrupt selection factors are applied [25]. In arable fields with repeated weed management in sequential cropping seasons (same herbicide, soil tillage, crop species), associations among weed species are expected to be valid at a higher degree compared to natural environments, since unfavored species are often eliminated from the area by the weed control techniques. For instance, following repeated application of a single herbicide, those weeds that still remain into the field are most probably those who present, as a common feature, the ability to tolerate that given herbicide, be it tolerant or resistant to that herbicide [58].

In weed science, an overall comprehension regarding plant associations is usually ignored, but its importance lays on two aspects: (1) weed species with positive association among them may answer better to environmental stresses as temperature and water shortage or excess [26, 40, 54], thus, associated plants are most prone to survive, reproduce, and increase its frequency into the community as they work together and (2) the understanding of the association among weed species in arable fields would make it possible to elaborate optimized control plans, be it chemical or cultural, which are efficient over a wider range of weed species at the same time since the technician previously knows they occur together. With the characterization of weed clusters in arable fields, it would be possible to estimate the appearance of weed species into the previously characterized clusters, even before its emergence, by observing the weed species already present and comparing to the usual cluster for that given crop and management.

Thus, understanding the association among weed species in arable fields would ultimately subsidize the development of sustainable techniques for weed control, including optimized herbicide recommendations. The limitation of its application, however, is that clusters would have to be defined for every combination of crop species (soybeans, rice, maize…), cropping system (direct seeding, water seeding, conventional tillage…), and environmental conditions (mainly based on edaphoclimatic characteristics).

Several methods are available in the literature related to plant ecology to assess plant associations in natural environments [25, 26, 40], where low levels of stress and disturbance are usually present [40, 54], and the climax of the vegetation may be most wide and dynamic. Climax is roughly defined as the final and relatively permanent condition of species occurrence in a given environment, as function of climate and soil characteristics [9]. In arable fields, the vegetation climax is heavily biased by crop management; thus, the weed climax tends to be narrower compared to the observed in natural environments, with probable lower degree of uncertainties in weed cluster characterization.

In the present chapter, we aim only to use the ecological approach of plant association as a tool for the weed science, so part of the methodologies available for detailed ecological studies, for instance as presented by Braun-Blanquet [9] and Barbour et al. [26], will not be covered in the present text. The basic steps to achieve a relatively complete characterization of association among weed species, as will be discussed in the following sections, are presented in **Figure 6**.

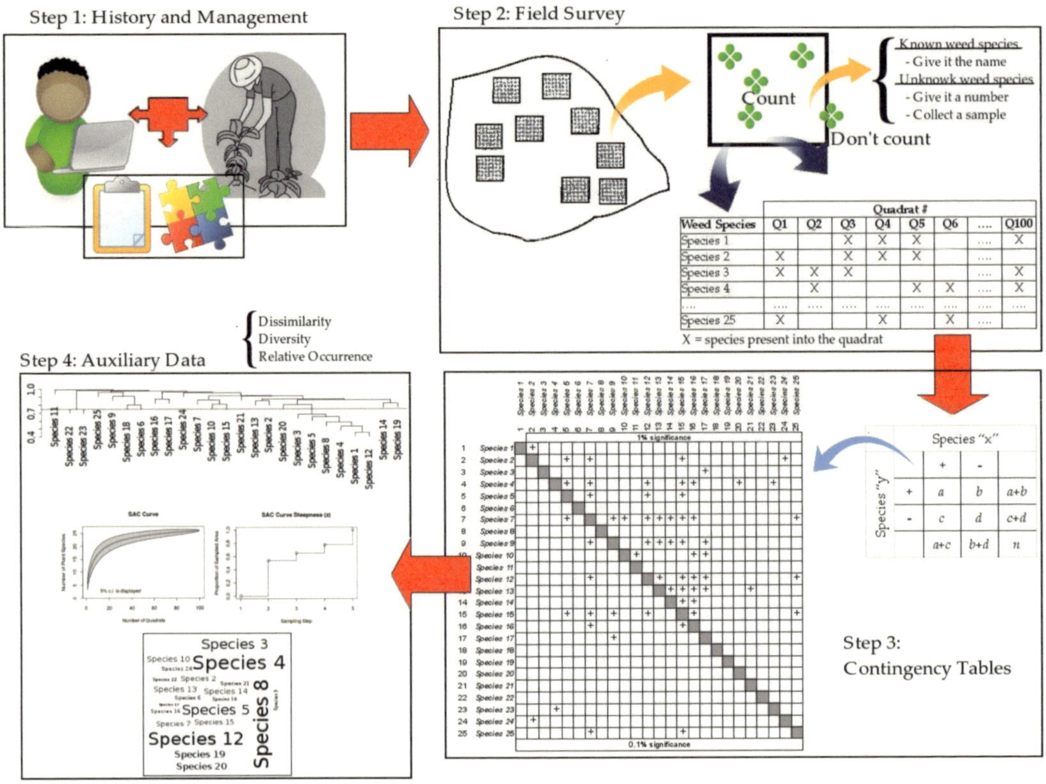

Figure 6. Basic steps to achieve a relatively complete characterization of association among weed species.

3.1. History and management of the area

The field survey for plant association, in arable fields, should be started only after an overview of the area has already been obtained from those who know the field very well. The first step in a study of plant association in arable fields includes obtaining data about the history of the area. Suitable weed management programs include extensive field scout for identifying weed populations and its seeds [59], as well as what growth stage the weeds are in Ref. [54]; talking to the farmer and his field workers will also supply valuable information about the predominance of weed species in the preceding years. Other information that should be obtained is the history about soil tillage, liming, and fertilization, as this information is needed to understand the biological nature of the predominant weed species in the area.

The main point, however, is the history of the herbicides previously applied to the field—those with long residual effect may heavily select weeds which are less susceptible to it [54]; the same may occur with frequent applications of non-residual herbicides [21]. The last 3 years of herbicide application, at least, should be known to the researcher. Perennial and long-term crops, as fruit trees or sugarcane, may mean that longer residual herbicides could have been applied, and in that case, the species associations are valid only under those or similar conditions, as in the absence of the residual effect of herbicides, other weed species would occur in that crop, at that location, and plant clusters would most probably be different.

3.2. Contingency tables

After the survey about the history and predominant management of the area is concluded, the second step in the determination of plant associations is a field survey by launching random quadrats with fixed size into the area. Methods for optimizing quadrat distribution in the field survey are available in ecology books [25, 26, 30, 37]. In general, sampling 100 quadrats should be enough for a reliable survey [26] in average size fields, although for plant association, both the correct quadrat geometrical form and size are of great importance. The optimal geometrical form for the quadrat is round or square [40], as it will reduce the total perimeter of the quadrat to the minimum and thus help reduce the error associated with the observer deciding if an individual of a rare weed species is in or out of the quadrat [26].

Quadrat size, however, is much more important than quadrat form as the data are of frequency type; correct quadrat size is preponderant as the X^2-test will be used to decide if some specific association between two weed species occurs more often than would be expected by chance [26]. Quadrat sizes which are appropriate to the study of small plants as arable weeds are of 10 × 10 or 25 × 25 cm [40]. As bigger the weeds to be studied, as bigger should be the quadrat size. This makes sense since as bigger the plant individual, as bigger its circumference of influence over the nearby vegetation. Supposing a larger-than-advised quadrat is used for the study, there will be an unreal increase in the number of reported associations by the X^2-test.

Another point to be discussed is that the X^2-test should not be used "as is" when any of the expected values is less than "5". In this case, Fisher's exact test would be chosen [40] but as this test is also involved in a series of statistical controversies, the advice is to use a great number of sampling points per area (at least 100) and to apply the Yates' correction to the X^2-test, what can be done automatically by most up-to-date statistical softwares as "R". This should be enough for the weed science.

For each sampling point, all plant species rooted into the quadrat should be identified and recorded; there is no need to count the number of individuals per species or to assess its dry mass. Plant species which are not known at the time of the evaluation should be identified by a number and have a sample collected for posterior identification by a plant taxonomist (**Figure 6**).

The plant species should be listed by sampled quadrat and compared in pairs, the data being organized in 2 × 2 contingency tables, as follows [40]:

		Species "x"		
		+	-	
Species "y"	+	*a*	*b*	*a*+*b*
	-	*c*	*d*	*c*+*d*
		a+*c*	*b*+*d*	*n*

where: a = number of quadrats containing both species; b = number of quadrats containing only species "y"; c = number of quadrats containing only species "x"; d = number of quadrats with both species absent; and n = total number of quadrats in the contingency table

The association between plant species is estimated by the chi-square (X^2) test, usually at 5 or 1% significance, simply represented by the formula [40]:

$$X_T^2 = \frac{(\text{obs} - \exp)^2}{\exp} \tag{16}$$

where: X_T^2= traditional chi square estimation; obs = observed values for species occurrence; and exp = expected values for species occurrence.

The expected values for each pair of occurrence in the 2 × 2 contingency tables are estimated by using the observed values from field sampling, as follows (**Table 3**).

As the association analysis uses 2 × 2 contingency tables, there is only one degree of freedom for the X^2 test. Besides, by anticipating that rare species (whose expected frequency is less than "5") may be reported, it is recommended to use Yates' correction for the X^2; thus, the following formula should be adopted as replacement for the traditional X^2 test, which automatically applies Yates' correction [40]:

$$X^2 = \frac{n[(|ad - bc|) - 0.5n]^2}{(a+b)(c+d)(a+c)(b+d)} \tag{17}$$

where: X^2 = chi-square estimation with Yates' correction; a = number of quadrats containing both species; b = number of quadrats containing only species "y"; c = number of quadrats containing only species "x"; d = number of quadrats with both species absent; and n = total number of quadrats in the contingency table.

The results of the calculated X^2 should be compared to the respective tables at one degree of freedom, and the results presented as a chi-square matrix crossing all species in pairs (**Figure 7**). The results indicate no association with a given pair of species when the probability in the chi-square tables, at the given number of degrees of freedom, is higher than the

Symbol/description		**Expected quadrats**[1]
a	"x" and "y" present	((a + b)/n) × (a + c) = K
b	"y" present	(a+b) – K = L
c	"x" present	(a+c) – K = M
d	none present	n – (K + L + M)

[1]The number of expected quadrats should be estimated by using the observed field values.
Source: adapted from Barbour [26].

Table 3. Contingency table estimators of expected values of occurrence for analysis of association between two species, "x" and "y".

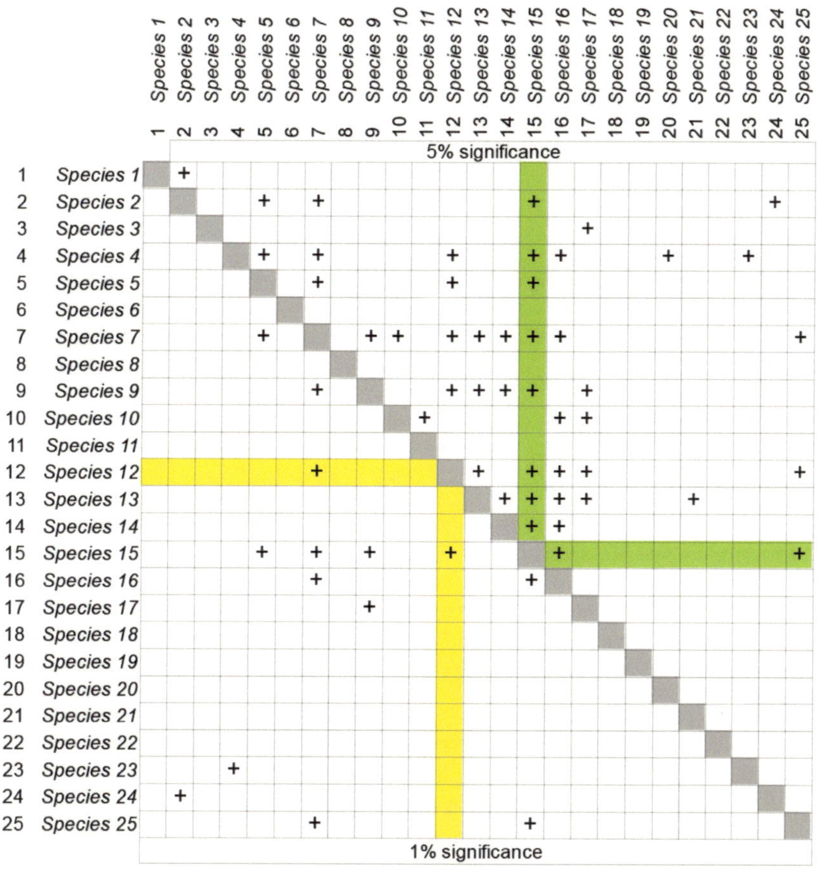

Figure 7. An example of an association matrix involving 25 weed species, at both 5 and 1% probability, according to the X^2-test. For reference, the association trees for species 15 (at 5% probability) and species 12 (at 1% probability) are highlighted in green and yellow, respectively. In this example, all associations are positive.

established significance level (usually 5 or 1%). The association is considered significant, as positive or negative association between two species, when the *observed values* of occurrence for a given pair of species are significantly higher or lower than the *expected values*, respectively [26]. An example of association matrix is supplied in **Figure 7**.

A dissimilarity between areas, Eqs. (14) and (15), is also useful for studies of plant associations. Similarly, diversity for intra-characterization of the areas may be applied to plant association studies in the same way they are applied for pure phytosociological studies (Eqs. (8) and (9)).

The relative occurrence of species and botanical families in the sampled area may be determined by its frequency of occurrence, usually considering the number of quadrats in which a given species is reported, related to the total number of quadrats, disregarding the number of individuals per quadrat:

$$Fr = 100\frac{f}{F} \quad (18)$$

where: *Fr* = frequency of occurrence for the given species; *f* = number of quadrats where the given species was present; and *F* = total number of sampled quadrats.

The frequency may be presented in several ways, but wordclouds make it easy to be understood. In wordclouds, the font size used to write the name of each species or family is proportional to their respective values of frequency (**Figure 6**).

4. Objections to the phytosociological method and application of the theory

Although being used for some years as a tool for the weed science, phytosociological surveys applied to arable fields have its drawbacks. As these methods were originally designed to describe natural environments, usually free from heavy anthropogenic effect, adaptations were needed for the agricultural context where the current flora present into the field is usually and mostly a result of the last cropping season's management (soil tillage system, fertilization levels, and herbicides applied, among other factors).

The main adaptations were (1) to establish the basic five steps for a reasonably complete phytosociological analysis, as described in the present text (overall infestation, phytosociological tables, diversity, similarity, and association); (2) to suggest and give preference to formulas which are less impacted by the most preponderant factors which could distort the phytosociological analysis, mainly for diversity and similarity; and (3) to use the method not only directly to the current flora into a given area but also to its seedbank through a germination study into controlled environment, as suggested by Concenço [60], and later comparing both studies (surface and seedbank samplings).

Another issue in the application of the method is its difficulty for both data collection in the field and its processing into the office, compared to what the researchers are familiar to analyze. Most weed science researchers usually adopt the visual method of evaluation for quantifying the occurrence of weeds into a given arable field, but this information is as easy as vague; it consists in taking note of the percentage of occurrence of each weed species into the field or alternatively—mainly following a herbicide application—evaluating the percentage of weed control some days after herbicide application. This method, although traditional and easy, does not supply at all information regarding the long-term behavior of weeds into the evaluated fields or its trend of occurrence for the next cropping seasons.

Another difficulty in applying the phytosociological methods for weed surveys is probably to convince the established weed science researchers to shift from the traditional evaluation methods (based on percentage of weed occurrence and control) to the phytosociological scope. The literature, however, proves that the adoption of such methods is highly positive for the sustainability of herbicide recommendations and weed management in the long term. One of the first Brazilian studies to apply the phytosociological method to the weed science, although in simple terms, was conducted by Carvalho and Pitelli [61]. Later, studies by Jakelaitis [62], Tuffi-Santos [63], Adegas [64], and several others adopted with success the phytosociological method for studies in weed science.

Although the use of phytosociological methods in the weed science is not new, the set of methods adopted is not standardized and ranges from basic to complex and from suitable to nearly unsuitable, depending on the paper. This makes almost impossible to compare studies conducted by different researchers as formulas and procedures are unlikely to be equivalent. The present chapter, however, partly intends to standardize the methods and its application.

5. Future insights

Weed science researchers will soon note that the traditional way of evaluating weed occurrence, infestation, or severity needs to move from a passive and subjective visually based assessment to most data-based decisions, and the phytosociology tends to be consolidated as the preponderant tool in this new universe of the weed science.

The difficulty in data collection for the phytosociological methods is still to be solved, but in the next few years, technologies such as GPS-driven drones with infrared imaging ability may be able to make data collection easier. Regarding data processing, the office work may still be an issue, but there are specific scripts for statistical softwares which could make the task of processing and interpreting the data easier, as the one published by Concenço [65], which makes possible to automatize phytosociological data processing into the statistical environment "R". This script, unfortunately, does not process the section of plant associations in its current version but is still a valuable tool that is freely available and adaptable.

Finally, an automatized integration from data collection into the field by GPS-driven drones, its transference to office and automatic processing by phytosociology software would provide farmers and technicians valuable tables and graphs for supporting both immediate and long-term decision-making in weed management.

6. Final considerations

This chapter discussed how some elements of phytosociology in ecology and botany can be used into the weed science as a tool for several inferences in arable fields. This is important to support recommendations for good agricultural practices while keeping up with biological conservation.

While choosing a sampling methodology for population studies in weed science, two questions should be taken into consideration: (1) *"what should be known?"* and (2) *"what will be done with the information?"* The first question addresses to the main type of information to be collected: richness or abundance. The second responds to the very purpose of sampling: biodiversity, evaluation of weed management plans, or studies in biology or ecology. By answering these two questions, one can design and choose a sampling methodology according to the information needs.

Weed relationships with edaphoclimatic traits show that the weed community is sensitive to variations in pH, water, temperature, and other resources and conditions. Each weed population is mostly competitive and dominant in those locations that meet particular conditions, and this would allow weeds to be used also as bioindicators as well as help understand the long-term dynamics of weed communities.

It is important to evaluate other sampling methods in order to know the sensitivity and accuracy of such alternatives in comparison with the ones presented here for distinct sampling objectives. One should ever think that as more species coexist in an association, and as greater sized are the plants, as bigger should be the minimum area to be sampled.

Phytosociological surveys are useful as tools to shed light on the dynamics of weed species and their interactions in arable fields. The methods, however, are the most diverse as several indices and coefficients are available, depending on the literature used as a reference by a given author. Basic care should be taken, however, when sampling and describing the plant community as well.

The following sequence of steps is proposed as suitable for phytosociological studies: (1) overall infestation; (2) phytosociological tables and/or graphs; (3) intra-characterization by diversity coefficients; (4) inter-characterization and area grouping by multivariate analysis; and (5) weeds association through contingency tables by means of the chi-square test. Other ways for presenting data should still be suitable, depending on the nature of the environment to be studied—arable fields in this case.

Literature is not clear about a set of methods for phytosociological studies, and one will hardly be able to find all the information and equations into the same source. Even classical references miss some aspects of phytosociological surveys, and some papers were published by using an unsuitable set of ecological methods to describe the weed community.

In the present chapter, a summary of methods was made in order to assist weed science researchers through their first steps into the realm of phytosociology.

Author details

Germani Concenço[1]*, Patrícia Menegaz de Farias[2], Néstor Fabio Alzate Quintero[3], Fábio Schreiber[4], Leandro Galon[5], Michely Tomazi[1], Ivana Santos Moisinho[4], Mariane Camponogara Coradini[4], William Christofari Ceolin[4] and André Andres[1]

*Address all correspondence to: germani.concenco@embrapa.br

1 Embrapa Temperate Agriculture, Pelotas, Brazil

2 University of South Santa Catarina—Unisul, Tubarão, Brazil

3 Ecology Institute—Inecol, Xalapa, Mexico

4 Federal University of Pelotas, Brazil

5 Federal University of the Southern Border, Erechim, Brazil

References

[1] Westhoff V, Van der Maarel E. The Braun-Blanquet approach. In: Whittaker RH, editor. Ordination and Classification of Plant Communities. 2nd ed. The Hague: Dr. W. Junk; 1978:287–399

[2] Pott R. Phytosociology: A modern geobotanical method. Plant Biosystems. 2011;**145**:9–18

[3] Lévêque C. Ecology from Ecosystem to Biosphere. Enfield, NH: Science Publishers Inc.; 2003. p. 490

[4] Ricklefs RE. A comprehensive framework for global patterns in biodiversity. Ecology Letters. 2004;**7**:1–15

[5] Hubbell SP. The Unified Neutral Theory of Biodiversity and Biogeography. Princeton, NJ: Princeton University Press; 2001. p. 392

[6] Holyoak M, Leibold MA, Holt RD. Metacommunities: Spatial Dynamics and Ecological Communities. Chicago, IL: University of Chicago Press; 2005

[7] Weber HE, Moravec J, Theurillat JP. International code of phytosociological nomenclature. 3rd ed. Journal of Vegetation Science. 2000;**11**:739–768

[8] Blasi C, Frondoni R. Modern perspectives for plant sociology: The case of ecological land classification and the ecoregions of Italy. Plant Biosystems. 2011;**145**:30–37

[9] Braun-Blanquet J. L'école phytosociologique zuricho-montpelliéraine et la S.I.G.M.A. Vegetatio. 1968;**16**:2–78

[10] Pignatti S, Oberdorfer E, Schaminée JHJ, Westhoff V. On the concept of vegetation class in phytosociology. Journal of Vegetation Science. 1995;**6**:143–152

[11] Oosting HJ. The Study of Plant Communities: An Introduction to Plant Ecology. San Francisco, California: Freman; 1956. p. 440

[12] Harper JL. Population Biology of Plants. Londres: Academic Press; 1977. p. 892

[13] Garnier E, Navas ML. A trait-based approach to comparative functional plant ecology: Concepts, methods and applications for agroecology. A review. Agronomy and Sustainable Development. 2012;**32**:365–399

[14] Concenço G, Tomazi M, Correia IVT, Santos SA, Galon L. Phytosociological surveys: Tools for weed science? Planta Daninha. 2013;**31**:469–482. DOI: 10.1590/S0100-83582013000200025

[15] Roberts DW. A dynamical systems perspective on vegetation theory. Vegetation. 1989;**69**: 27–33

[16] Roberts DW. Fuzzy systems vegetation theory. Vegetation. 1989;**83**:71–80

[17] Ostermann OP. The need for management of nature conservation sites designated under Natura 2000. Journal of Applied Ecology. 1998;**35**:968–973

[18] Kratochwil A, Schwabe A. O˙kologie der Lebensgemeinschaften. Stuttgart: Ulmer; 2001

[19] Santos WF, et al. Weed phytosociological and floristic survey in agricultural areas of Southwestern Goiás region. Planta Daninha. 2016;**34**(1):65–80

[20] Grime JP. Plant Strategies and Vegetation Process. New York: John Wiley & Sons; 1979. p. 209

[21] Pitelli RA, Pitelli RLCM. Biologia e ecofisiologia das plantas daninhas. In: Vargas L, Roman ES, editors. Manual de Manejo e Controle de Plantas Daninhas. Bento Gonçalves: Embrapa Uva e Vinho; 2004. pp. 29–56

[22] Biondi E, Feoli F, Zuccarello V. Modelling environmental responses of plant associations: A review of some critical concepts in vegetation study. Critical Review Plant Science. 2004;**23**:149–156

[23] Ge´hu JM, Rivas-Martinez S. Notions fondamentales de phytosociologie. Ber Int Symp Int Vereinigung Vegetationsk. 1981;**1980**:5–30

[24] Begon M. Ecologia: De Individuos a Ecossistemas. Porto Alegre: Artmed Editora; 2009. p. 740

[25] Gurevitch J, Scheiner SM, Fox GA. Plant Ecology. Porto Alegre: Artmed; 2009. p. 574

[26] Barbour MG, Burk JH, Pitts WD, Gilliam FS, Schwartz MW. Terrestrial Plant Ecology. 3rd ed. Menlo Park, CA: Benjamin-Cummings; 1998. p. 688

[27] Clements FE. Plant Succession: Analysis of the Development of Vegetation. Publication no. 242. Washington, DC: Carnegie Institute of Washington; 1916

[28] Capelo J. Conceitos e métodos da Fitossociologia. Formulação contemporânea e métodos numéricos de análise da vegetação. Estação Florestal Nacional, Sociedade Portuguesa de Ciências Florestais, Oeiras; 2003. p. 107

[29] Gleason HA. The individualistic concept of the plant association. Bulletin of the Torrey Botanical Club. 1926;**53**(1):7–26

[30] Pandeya SC, Puri GS, Singh JS. Research Methods in Plant Ecology. New York: Asia Publishing House; 1968. p. 272

[31] Stohlgren TJ. Measuring Plant Diversity: Lessons from the Field. New York: Oxford University Press; 2007. p. 390

[32] Krebs CJ. Ecological Methodology. University of British Columbia, Vancouver, Canada; 1989. p. 654

[33] Terradas J. Ecología de la Vegetación. Ed. Barcelona, Spain, Omega; 2001. p. 760

[34] Whittaker RH. Classification of plant communities. Botanical Review. 1962;**28**:1–239

[35] Labrada R, Caseley JC, Parker C. Manejo de Malezas para Países en Desarrollo. FAO (Food and Agriculture Organization). Roma; 1996. p. 317

[36] FAO. Consulta de Expertos en Ecología y Manejo de Malezas. División de Producción y Protección Vegetal Organización de la Alimentación y la Agricultura de las Naciones Unidas. Roma: FAO; 1997. p. 60

[37] Booth B, Murphy S, Swanton C. Weed Ecology in Natural and Agricultural Systems. UK: CAB International; 2003. p. 303

[38] Marshall EJP. Field-scale estimates of grass weed populations in arable land. Weed Research. 1988;**28**(3):191–198

[39] Rew LJ, Cousens RD. Spatial distribution of weeds in arable crops: Are current sampling and analytical methods appropriate? Weed Research. 2001;**41**(1):1–18

[40] Moore PD, Chapman SB. Methods in Plant Ecology. Oxford: Blackwell; 1986. p. 589

[41] Goodall DW. A new method for the analysis of spatial pattern by random pairing of quadrats. Vegetatio. 1974;**29**:135–146

[42] Mueller-Dombois D, Ellenberg H. Aims and Methods of Vegetation Ecology. Wiley; 1974. p. 547

[43] Margalef DR. Information theory in ecology. General Systematics. 1958;**3**:36–71

[44] McManus JW, Pauly D. Measuring ecological stress: Variations on a theme by R. M. Warwick. Marine Biology. 1990;**106**(2)305–308

[45] Concenço G, Andres A, Teló GM, Martins MB, Moisinho MB. Phytosociological characterization of weeds as a function of residual herbicides applied to rice grown under sprinkler irrigation. Experimental Agriculture. 2017;**53**:1–12. DOI: 10.1017/S0014479717000059

[46] Gauch HG Jr. Multivariate Analysis in Community Ecology. Cambridge, UK: Cambridge University Press; 1982. p. 298

[47] Pielou EC. The Interpretation of Ecological Data: Aprimer on Classification and Ordination. New York: J.Wiley & Sons; 1984. p. 263

[48] Matteucci SD, Colma A. Metodología para el estudio de la vegetación. The General Secretariat of the Organization of American States, Washington D.C., USA; 1982. p. 168

[49] Moreno CE. Métodos para medir la biodiversidad. Vol. I. Zaragoza, España: M&T-Manuales y Tesis SEA; 2001. p. 84

[50] Concenço G, Silva CJ, Staut LA, Pontes CS, Laurindo LCAS, Souza NCDS. Weeds occurrence in areas submitted to distinct winter crops. Planta Daninha. 2012;**30**:747–755. DOI: 10.1590/S0100-83582012000400008

[51] Sneath PH, Sokal RR. Numerical Taxonomy: The Principles and Practice of Numerical Classification. San Francisco: W. H. Freeman; 1973

[52] Sokal RR, Rohlf FJ. The comparison of dendrograms by objective methods. Taxon. 1962;**11**(1):33–40

[53] Agostinetto D, Galon L, Moraes PVD, Rigoli RP, Tironi SP, Panozzo LE. Competitividade relativa entre cultivares de arroz irrigado e biótipo de capim-arroz (*Echinochloa* spp.). Planta Daninha. 2008;**26**:757–766. DOI: 10.1590/S0100-83582008000400007

[54] Radosevich S, Holt J, Ghersa C. Ecology of Weeds and Invasive Plants: Relationship to Agriculture and Natural Resource Management. New York: Wiley; 2007. DOI: 10.1002/9780470168943

[55] Stone LF, Silveira PM. Competition limits of grain yield components of common bean cultivar Perola. Bioscience Journal. 2008;**24**(2):83–88

[56] Galon L, Concenço G, Ferreira EA, Silva AF, Ferreira FA, Noldin JA, Freitas MAM. Competition between rice plants and *Echinochloa* spp. biotypes resistant or susceptible to quinclorac. Planta Daninha. 2009;**27**(4):701–709

[57] Vida FBP, Laca EA, Mackill DJ, Fernandes GM, Fischer AJ. Relating rice traits to weed competitiveness and yield: A path analysis. Weed Science. 2006;**54**(6):1122–1131

[58] Agostinetto D, Vargas L. Weed Resistance to Herbicides in Brazil. Pelotas: UFPel; 2014. p. 398

[59] Zimdahl RL. Fundamentals of Weed Science. San Diego: Harcourt Brace; 1993. p. 450

[60] Concenço G, Salton JC, Brevilieri RC, Mendes PB, Secretti ML. Soil seed bank of plant species as a function of long-term soil management and sampled depth. Planta Daninha. 2011;**29**(4):725–736. DOI: 10.1590/S0100-83582011000400002

[61] Carvalho SL, Pitelli RA. Survey and phytosociological analysis of the main weed species in pastures at the region of Selviria (MS). Planta Daninha. 1992;**10**(1/2):25–32

[62] Jakelaitis A, Ferreira LR, Silva AA, Agnes EL, Miranda GV, Machado AFL. Weed population dynamics under different corn and bean production systems. Planta Daninha. 2003;**21**(1):71–79

[63] Tuffi-Santos LD, Santos IC, Oliveira CH, Santos MV, Ferreira FA, Queiroz DS. Phytosociological assessment of degraded pastures under flooded low land conditions. Planta Daninha. 2004;**22**(3):343–349

[64] Adegas FS, Oliveira MF, Vieira O, Prete CE, Gazziero DLP, Voll E. Phytosociological survey of weeds in sunflower crop. Planta Daninha. 2010;**28**(4):705–716

[65] Concenço G. Phytosociological Analysis of Weeds: Method Applied to the "R" Environment. Dourados: Embrapa; 2015. p. 33. Available from: http://ainfo.cnptia.embrapa.br/digital/bitstream/item/131016/1/DOC-129-2015-completo-anexos.pdf

Chapter 8

Ecological Response to Global Change: Changes in C:N:P Stoichiometry in Environmental Adaptations of Plants

Renato de Mello Prado and Gilmara Pereira da Silva

Additional information is available at the end of the chapter

http://dx.doi.org/10.5772/intechopen.69246

Abstract

This review aims to discuss the state of the art of the stoichiometric ratio of foliar nutrients and their impact on adaptive mechanisms of plants to environmental change. Plant stoichiometry is an excellent way to study the multiple ratios across the nutrients in plants and their ecological interactions with the environment. It plays an important role in clarifying the responses of plants to various changes and their adaptation to different environments. However, anthropic activity can change the stoichiometric ratios of plants. In recent decades, anthropic activities have altered the cycle of nitrogen (N), phosphorous (P) and carbon (C) in plants. This is due to excessive fertilizer application, increased global warming and increased atmospheric CO_2 emissions, which can quickly limit the increase of production in plants, as they affect the process of acclimatization, which involves a series of changes in plant metabolism at different levels of organization (molecular, biochemical, anatomical and morphological). In this sense, in this new scenario of changes, new plant responses to stoichiometric changes and adaptive processes in the ecosystem have to be reviewed.

Keywords: plant species, multiple ratios, adverse habitats, ecological processes, ecosystem

1. Introduction

The increase in warming and droughts and the high concentrations of atmospheric CO_2 can change the contents and the stoichiometry of nitrogen (N) and phosphorous (P) in plants [1], and they can have an indirect impact on soil and nutrient availability. This increase in high

CO_2 concentrations induces changes in plants, especially in C3 plants, with an increase in the uptake of carbon (C), which may lead to a reduction of transpiration [2], because CO_2 absorption promotes stomata closure [3], which may limit the ability of plants to assimilate N [4].

The high absorption of CO_2 may also lead to a gradual limitation of nutrients that can quickly limit the increase in plant production [5], because it affects the process of acclimatization that involves a series of changes in plant metabolism at different levels of organization (molecular, biochemical, anatomical and morphological) [6].

In recent decades, anthropic activities have altered the P cycle; excessive doses of fertilizers are being used, thus inducing an increase in the input of this nutrient into terrestrial and aquatic ecosystems [7, 8]. Increased application of P may alter the balance between C, N and P in plants, and thus change the C:N:P stoichiometry ratios [8] and reduce the C:P ratio in plant tissues [9, 10]. Another concern is the change in N and P cycles, which can cause several consequences to the environment [11–13].

In this scenario, environmental responses of plants to global changes have a negative character with future losses to food production worldwide. Therefore, it is necessary to recognize the new stoichiometry (C:N:P ratios) that occurs in plants in this new scenario in order to try to identify a plant-environment interaction that may allow an increase in food production and that will allow greater food security in the future.

The interactions that occur between elements are complex and their effects reflect the mineral composition of plants. An alternative to study the multiple ratios between elements in a plant is to focus on stoichiometric ratios that are considered to be an important biological indicator for elucidating plant responses to various changes and their adaptation to different environments [14].

Moreover, the study of plant stoichiometry can influence ecological processes, and thus modulate the structure and function of the ecosystem [15, 16]. It can also effectively indicate changes in C, N and P cycles [17].

The carbon (C):nitrogen (N):phosphorus (P) ratio is one of the most investigated topics in stoichiometry, because N and P limit plant growth and C is the structural basis of plants: they account for 50% of plant dry mass [18].

These elements are strongly linked to the biochemical functioning of plants. P is an important element in the production of ribosomes; it is involved in the synthesis of proteins containing N and C. There are, therefore, fundamental biochemical reasons for using these elements in appropriate proportions [19].

In plants, C:N and C:P ratios represent the ability of photosynthetic fixation of C through N or P accumulation. Also, the N:P ratio can be used as an indicator to study plant nutrient limitation in adverse habitats [20].

Therefore, the proportions of leaf N and P in plant biomass can be an indicator of vegetation composition and nutrient limitation at the community level [21, 22]. An N:P < 14 ratio indicates N limitation, whereas an index >16 suggests P limitation [21]. An ideal N:P ratio is considered to be 10–20, on a mass basis [22].

In view of the above, the present chapter sought to study patterns and values and discuss the stoichiometric changes C:N:P occurring in plants in response to global changes and their implications in the adaptive mechanisms of plants to the environment.

2. Climatic effect

The climate exerts a strong control on plant growth and hence it influences plant stoichiometry. Changes in growth rate can be caused by changes in the availability of elements as a result of changes in temperature, latitude, drought and warming. Thus, one of the challenges in the future should elucidate the reasons and implications of this variability which may alter the success of resident plant species.

2.1. Latitude

Latitude is a climatic parameter that can influence stoichiometric ratios. In this scenario, three analyses of leaf N:P patterns indicated that the N:P ratio is approximately half when latitude in the Equator is 70° (**Figure 1**) [23, 24].

The reason for this trend can be explained by N concentrations (N:C) which are approximately constant for latitudes, increasing P concentrations as latitude changes. This is indicative of a trend in N:P [23].

A study that analysed foliar N:P ratio as a function of latitude showed that this ratio increases with temperature [25]. This increase in temperature towards the Equator occurs because P is

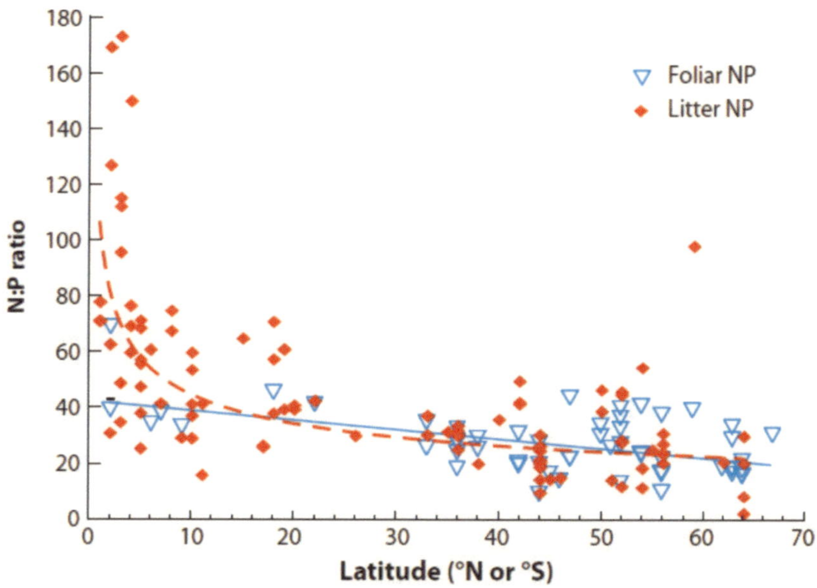

Figure 1. Variation in N:P (molar ratio) in foliage (open triangles) and litter (solid diamonds) as a function of latitude [23].

an important limiting nutrient in tropical soils and N is the main limiting nutrient in temperate regions and high-latitude soils.

2.2. Light

Differences in the exposure of plants to sunlight can also affect their stoichiometry. One study compared the N:P ratio of sunlight-exposed leaves and shade-exposed leaves of two species of *Quercus ilex* and *Quercus coccifera* plants in Spain [26]. In both species, the sunlight-exposed leaves had about twice as much concentration of P compared with the shade-exposed leaves, while there were minor differences in the concentrations of N. The P:N ratio was also higher in the sunlight-exposed leaves than in the shade-exposed ones.

This result indicates that the two plant species may show ability to adequately respond to changes in environmental factors by means of phenotypic plasticity, which is positively related to the ecological distribution of species.

2.3. Drought

Long periods of water stress often cause a reduction in plant growth [27], but plants respond with increased absorption of water and improved mechanisms for water use efficiency. The events caused by water stress initiate physiological responses in plants which often affect ecosystems and nutrient cycling [28, 29].

Mathematical models predict an increase in water deficit in various areas of the world. The effects of increased water deficit differ across ecosystems and species. In semi-arid areas, drought reduces the C:N ratio in the roots of the species *Quercus ilex* [30, 31].

In other plant species, drought increased the C:N and C:P ratios of leaves of shrubs and trees in the Mediterranean, as a result of protection mechanisms [32, 33] associated with the presence of leaves whose structure is drought-tolerant [34].

In moist temperate ecosystems, the C:N ratio can decrease moderately because plants increase the uptake of N and reduce their growth [35].

Thus, evidence suggests that drought tends to increase C:N ratios of photosynthetic tissue in semi-arid environments, but the effects are not so clear in moist ecosystems (**Figure 2**), in which drought may affect various aspects of plants.

In dry regions, the increase in C:N ratios can combine with increases in response to CO_2 concentrations, which suggests synergy that increases the C:N ratio (and, probably, the C:P ratio) and slows the N and P cycles, thus reducing the availability of N and P and their concentrations in the biomass [32, 36].

2.4. Warming

The increase in ambient temperature can increase the mineralization of organic C and, thus, increase the amount of atmospheric CO_2 [37]. This may explain why several studies have not detected an effect of high ambient temperatures on the C:N ratios of some plants (**Figure 2**).

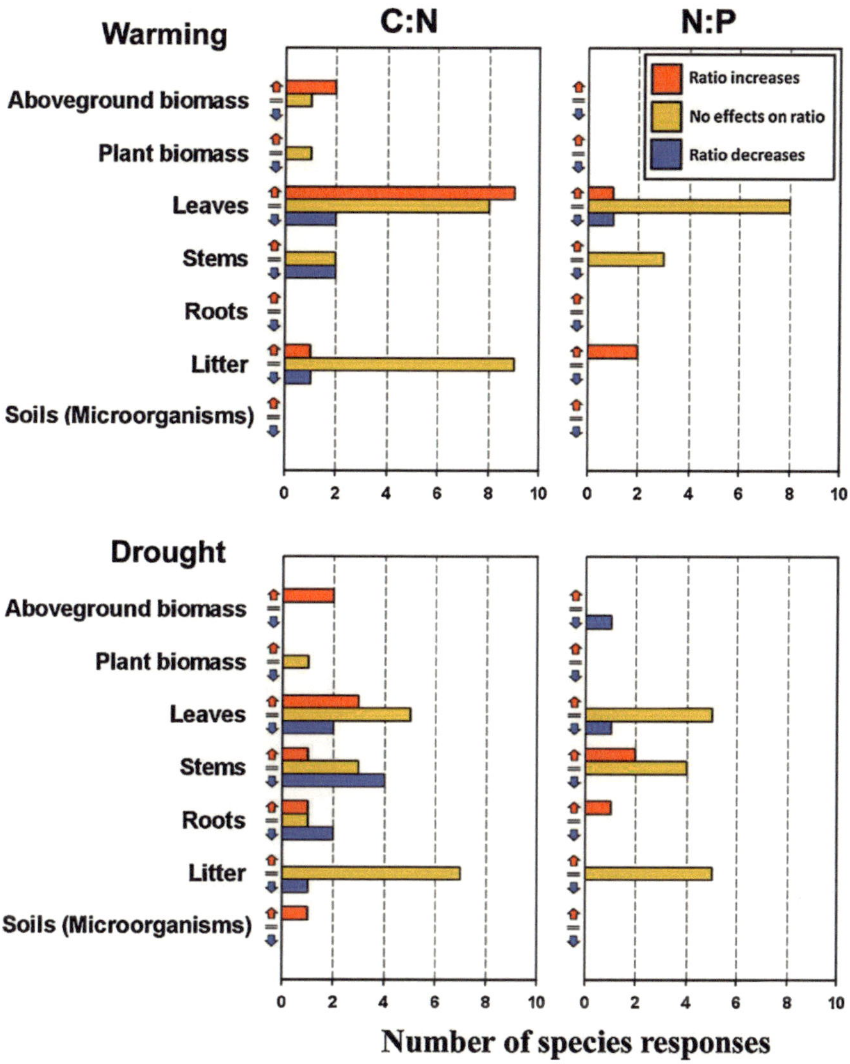

Figure 2. Reported increases, decreases and absence of change in C:N and N:P ratios in response to warming or drought [13].

Plant respiratory responses to warming may affect the availability of light, water and CO_2, and plant responses may differ across species and organs [38]. Thus, evidence suggests that the rise in temperature predicted by climate models will increase the C:N and C:P ratios of plants based on mechanisms of water stress resistance or water-use efficiency. This increase of C:N and C:P ratios caused by warming coincides in semi-arid regions.

In cold ecosystems that are not limited by water, the effects of warming on C:N ratios of plants are not well understood. However, some studies have shown that warming has changed the C:N ratios of plants, increasing production capacity and nutrient absorption of plants [39]. Other studies in pastures in cold regions have not found any effects [40] or have reported an increase in the C:N ratio associated with an effect of dilution by an increase in biomass production [41].

3. Variations in C:N:P ratios in plants

3.1. Growth rate and N:P stoichiometry in plants

The applicability of growth rate (GR) to plants has been attracting interest because leaves have high concentrations of nutrients (N and P). However, only a few experimental studies have assessed GR for particular plant species.

A study with seedlings of the species *Betula pendula* with P limitation showed a decrease in N:P ratios and high relative growth rates; however, plants with N limitation did not show this pattern [42], probably because of P storage under N limitation.

A study with 14 pine species grown with high levels of nutrients in a greenhouse [43] reported faster plant growth, which was correlated with nutrient concentrations and a decrease in the protein RNA.

Finally, when the researchers compared the seedlings for growth rates among the 14 species under high-nutrient conditions, they found no correlation with N:P ratios or the protein RNA [43].

The results of Ågren [42] and Matzek and Vitousek [43] suggest that the basic prediction of GR (a negative correlation between N:P and growth rate) may not be useful for plants when nutrients, especially P, are not limiting factors.

Thus, studies have concluded that although the vegetable protein:RNA ratio affects the speed and efficiency of growth; it does not determine, by itself, leaf N:P stoichiometry. Thus, it seems that the advances and understanding of interactions between N:P stoichiometry and growth require both additional studies and development of models that represent the potential storage of nutrients, especially P [44].

The correlation between N and P observed in leaves has been recently confirmed in other important plant organs [45]. A study using a high number of species [46] found that, as in leaves, N and P concentrations are correlated in roots, stems and in reproductive tissues.

High concentrations of P and low N:P ratios are linked with growth rate [14]. The effects of high CO_2 concentrations on plants and C:P ratios and, in particular, on N:P ratios, still are not clear [47] and deserve further study because N and P are essential in living systems and their relationship is associated with changes in the structure of the ecosystem [14]. In fact, it is important to increase our understanding of changes in the mineralization of N and P in plants and in the soil under high CO_2 concentrations because N and P are important in the composition of litter and soil decomposition rates [48]. A meta-analysis showed that added N significantly decreased the C:N ratio of photosynthetic tissues of woody plants ($P < 0.05$, $n = 25$) and herbaceous species ($P < 0.05$, $n = 6$). On average, N reduced the C:N ratios of photosynthetic tissues by 25% ($P < 0.05$, $n = 31$) (**Figure 3**).

In 20 of 36 species, the addition of N increased the N:P ratio of photosynthetic tissues; in 15 species, the N:P ratios were not changed and, in one species, the ratio decreased. The addition of N increased the N:P ratios of the photosynthetic tissues of woody plants ($P < 0.05$, $n = 10$) and herbaceous species ($P < 0.05$, $n = 12$) (**Figure 3**).

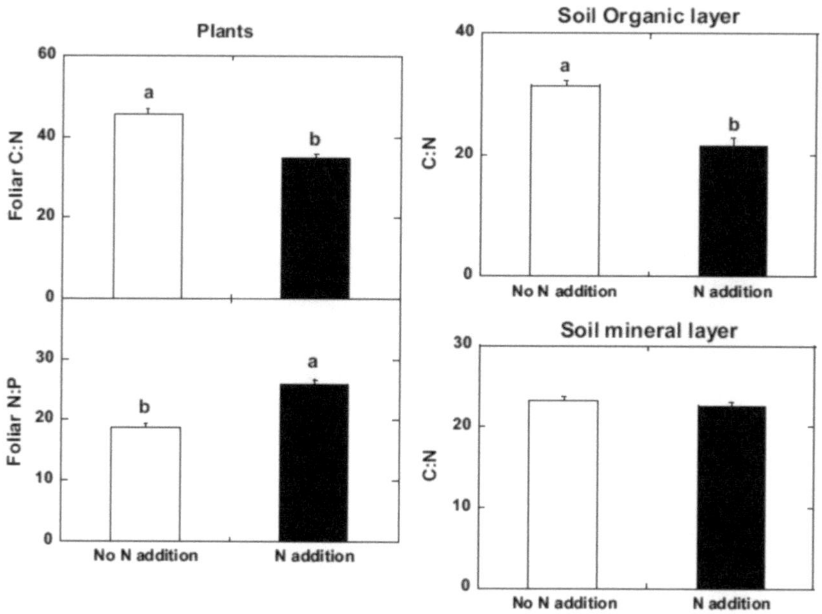

Figure 3. Leaf C:N and N:P ratios of plants growing in the environment and in the N-addition treatment. Metadata analysis of 31 different experimental results in the case of leaf C:N ratio and of 22 different experimental results in the case of leaf N:P ratio. Soil organic and inorganic C:N ratios under ambient conditions and in the N-addition treatment. The only studies that were taken into account were those that provided the mean (±S.E.) of leaf C:N and C:P ratios of plants, and soil C:N ratios growing under ambient conditions and in the N-addition treatment. Meta-analyses were made by using the MetaWin Package, which is based on the knowledge of control and treatment results (mean ± SD) in each study (considering each species being studied). Different letters indicate statistically different values ($P < 0.05$) [13].

3.2. Plant stoichiometry

Plant stoichiometric is a technique that allows investigating energy flow and cycling of materials in ecosystems [49], stoichiometric flexibility, physiological adjustment of C:N:P ratios which may improve plant performance in response to environmental changes [50]. Therefore, it is important to investigate the patterns of stoichiometric values and their flexibility within and among plant species [51].

However, this technique may present some disadvantages regarding the variability that occurs in plant C:N:P stoichiometry in several habitats and emerges from two interaction processes: (1) macro-scale constraints caused by specific geographic environment (climate and soil), and (2) fundamental physiological constraints resulting from growth, development, metabolism, phenology and life history [23].

Moreover, plant size, which changes as a result of seasonal development, may influence the rate indicated by the metabolic scale theory [52, 53], which in turn affects the stoichiometric ratios through metabolic changes [1].

Sampling time is another factor that may compromise the success of this technique, because sampling ranges from months to years, and the effects of organ size within a period of the year of the study are often not kept constant [25, 54].

Thus, a study developed to evaluate the C:N:P stoichiometric flexibility as of the date of sampling was developed by Zhang et al. [55] in field conditions in Mongolia. Three plant species were selected: *Leymus chinensis* (perennial C3 plant); *Cleistogenes squarrosa* (perennial C4 plant) and *Chenopodium glaucum* (annual C3 plant).

To study the effects of sampling date, 30 individual plants of each species were collected at 15-day intervals, from 10 July to 25 August 2006, for a total of four sampling dates.

The authors found that the C:N, C:P and N:P ratios in the leaf tissue increased over time compared with the study species, except for the species *Chenopodium glaucum* (**Figure 4a–c**) [55].

For the species *Leymus chinensis*, the C:N, C:P and N:P ratios were the highest among the three species and they increased over time, with the exception of the N:P ratio until the last date of

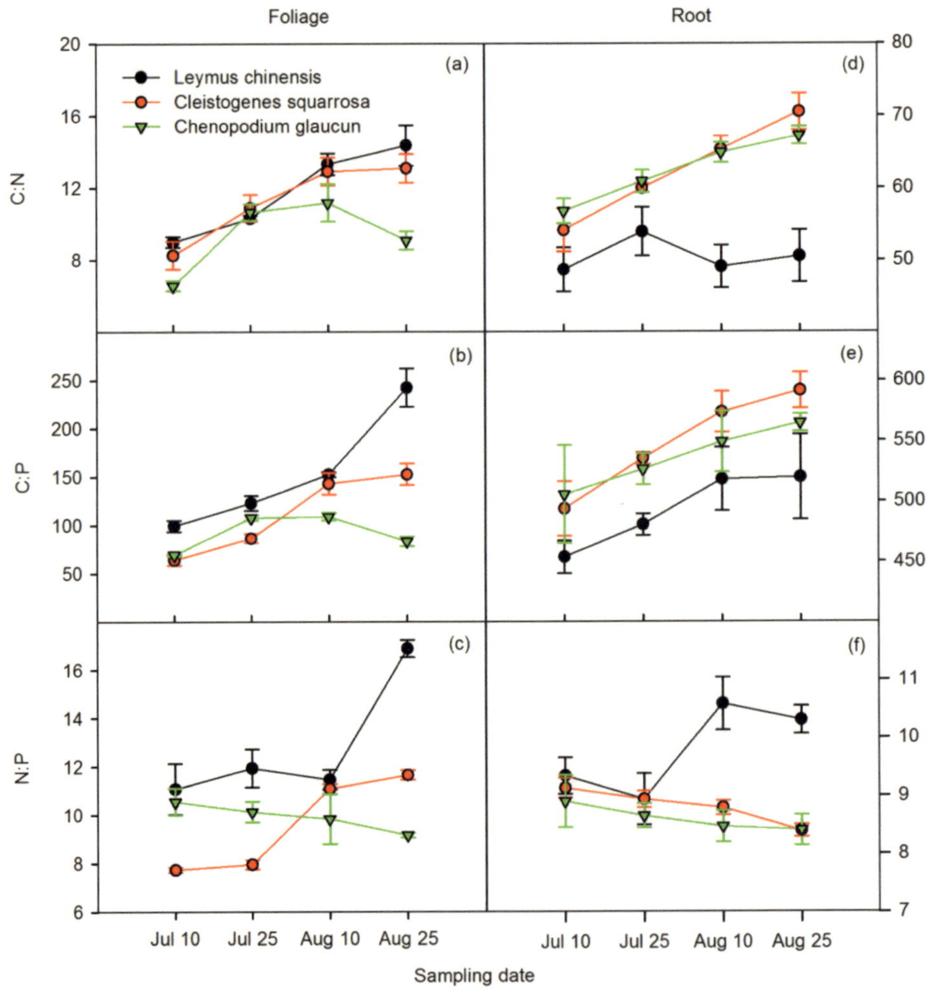

Figure 4. Change in C:N (a), C:P (b), N:P (c) ratios for leaf (left) and C:N (d), C:P (e), N:P (f) root (right) tissues over time for three grassland species in the sand culture study. Error of mean [49].

sampling (**Figure 4c**). However, for *Chenopodium glaucum*, the C:N and C:P ratios increased for the first two sampling dates and then decreased after 10 August 2006 (**Figure 4a** and **b**) [55].

Thus, the study suggests that leaf sampling at different times may influence the stoichiometric ratios of the plant, particularly C:N and C:P ratios of leaves [55].

In general, the C:N and C:P ratios of leaves increased with increasing sampling date within the study periods. This increase was probably driven by the increase in plant size (C content); as plants get older, the C-enriched material accumulates, which leads to a 'dilution' of N and P contents over time [56, 57].

Thus, over time, C:N and C:P ratios may increase because of reduced nutrient allocation to older leaves and to nutrient dilution as the leaf area and root systems increase over time [55].

C:N:P stoichiometric ratios in plants can also be altered depending on the application of beneficial elements in agriculture, e.g. silicon (Si).

In this scenario, a study was conducted in a greenhouse in Jaboticabal, São Paulo State, Brazil, in which a rice crop was combined with the application of Si sources (Nano silica and soluble silicon) and concentrations of Si (0, 605, 1210, 103 and 2420 g ha^{-1} Si, applied on the seeding furrow). They found that Si availability did not affect the C:N:P stoichiometric ratio in the shoot of rice plants, although there were higher stoichiometric C:N:P ratios in the concentration of 1210 and 2420 g ha^{-1} Si, when soluble silicon was used (**Table 1**) [58].

In this study, the stoichiometric ratio found refers to the average, excluding the panicle, and this probably resulted in the absence of more pronounced effects of the treatments applied [58].

There are strong associations between the absorption of Si, N and P, and a study on silicon sources and grass species emphasized that the responses varied according to the sources of silicon in use [59]. In an experiment with *Phragmites australis* [60], it is reported that Si availability can have significant effects on stoichiometric C:N:P ratios in different tissues (leaf blades, sheaths and stems).

Elements	**Treatments (g ha^{-1})**	**Stoichiometry**
C:N:P	Nano silica: 0	188:15:1
	Nano silica: 605	191:16:1
	Nano silica: 1210	183:15:1
	Nano silica: 2420	197:16:1
	Soluble silicon: 0	188:15:1
	Soluble silicon: 605	183:15:1
	Soluble silicon: 1210	199:17:1
	Soluble silicon: 2420	199:17:1

Table 1. Stoichiometry of nutrients affected by sources and doses of silicon applied in the seeding furrow of rice [58].

4. Global changes and plants: a perspective of stoichiometric scaling

Global changes can affect the stoichiometry of plants and of the ecosystem through changes in C:P and C:N ratios, which can alter food quality, affect the nutrient cycle, impoverish the nutrient composition of the ecosystem and increase the risk of extinction of species.

4.1. Atmospheric CO_2

There is strong evidence on the relationship between atmospheric CO_2 concentration and plant stoichiometry. It is expected that the increases of atmospheric CO_2 will stimulate the plant photosynthesis and, perhaps, growth and overall production.

As a result, there is potential for C sequestration in plant biomass as atmospheric CO_2 increases [61]. However, the length of plant growth in any location is probably influenced by the resources available in the soil, particularly N [62].

Atmospheric CO_2 fixation tends to increase the root/plant ratio [63] and leaf area [64], which will influence the C:N:P ratios of the entire plant and, ultimately, photosynthetic capacity [65].

At the molecular level, rubisco, a key photosynthetic enzyme, operates more efficiently at higher levels of CO_2 emissions (intracellular levels), especially in C3 plants [66], by minimizing the need for gene expression of the enzyme to compensate for the losses to photorespiration [67]. The resources (for example, N) which are not used to produce rubisco can then be diverted to increase production [68].

In general, a higher concentration of CO_2 should result in a greater C:N ratio in plant biomass and increases in plant size [69].

4.2. Global warming

Global warming will likely influence plant stoichiometry, plant species, community primary production through impacts on phenology and plant growth conditions [61].

However, these effects will be moderated by drought. For example, in the long term, warming with increasing drought conditions in the Amazon can induce massive changes in biomass carbon [62].

However, restrictions on the use of nutrients [45] and changes in development and the way plants share resources across the types of tissue [70] suggest changes in C concentrations on a large scale; they will also be accompanied by absolute changes in levels of soil nutrients.

4.3. Varying increases in supplementation with N and P

The majority of terrestrial ecosystems has historically been adapted to a natural limitation of key nutrients [71]. Combustion of fossil fuels, use of fertilizers, agricultural production of legumes [72], deforestation and changes in land [73] allowed for a large-scale duplication of input of biologically available N in ecosystems around the world. The anthropogenic effects of P in the biosphere appear to be even greater, because the cycle of this nutrient was amplified four times by human action [74].

In the short term, more availability of N and P can increase the productivity of plant species through a greater leaf area index [65] among other routes, and biomass [18]. In the long term, increases mediated by nutrient deposition in the soil can shape community composition differentially, changing the growth rate and the success of resident plant species [75].

5. Methodology

Data from the published literature on the ecological response to global change were collected: C:N:P stoichiometry changes in plant environmental adaptations. The information on the climatic effect of latitude was obtained from information on variation in N:P (molar ratio) in foliage (open triangles) and litter (solid diamonds) as a function of latitude [23].

Regarding the light climatic factor, data were selected and collected on differences in the exposure of plants to sunlight that can affect plant stoichiometry. This study was based on the work of Valladares et al. [26].

Regarding the dry climatic factor, the information collection was made considering that the effects of the water deficit differed between ecosystems and species, based on scientific studies [30–35].

For the climate heating factor, information on the effect of temperature, respiratory responses to heating and on cold ecosystems the effects of heating on the C:N ratios [38–41] were addressed.

The variations in the C:N:P ratios in plants were plotted from the collection of information on growth rate and N:P stoichiometry in plants that were based primarily on works of [42, 43], and plant stoichiometry in the information collected mainly from the work of [49, 50, 55, 58].

Finally, global and plant changes: a perspective of stoichiometric design, atmospheric CO_2 was studied; Global warming and variable increases in N and P supplementation, from the collection of relevant information, based on published scientific articles on the subject.

6. Conclusions and future perspectives

This chapter addressed the main issues regarding the ecological response to global change: C:N:P stoichiometry changes in plant environmental adaptations, based on recent scientific findings that can guide students and researchers in their studies and enable future research.

Plant stoichiometry is becoming an excellent measure to study the multiple ratios that occur between plant nutrients and their ecological interactions with the environment. It plays an important role in elucidating a plant's responses to various changes and adaptation to different environments. However, anthropic activity can change stoichiometric ratios of plants.

In recent decades, anthropic activities have altered the cycle of N, P and C, mainly with the use of high doses of fertilizers, increased global warming, droughts, increased atmospheric CO_2 emissions. This increase has altered stoichiometric relations among the nutrients in plant tissues and their availability, thus influencing the structure of ecosystems.

In this sense, future research needs to review the new responses of plants in relation to stoichiometric changes and processes adaptive to the ecosystem to this new scenario of changes.

Author details

Renato de Mello Prado and Gilmara Pereira da Silva*

*Address all correspondence to: gilmaraagronoma@gmail.com

Universidade Estadual Paulista Júlio de Mesquita Filho, Jaboticabal, SP, Brazil

References

[1] Rivas-Bach A, Sardines J, Pérez-Trujillo M, Estimate M, Pinellas J. Strong relationship between elemental stoichiometry and metabolome in plants. Proceedings of the National Academy of Sciences. 2012;**109**:4181-4186

[2] Del Pozo A, Perez P, Gutierrez D, Alonso A, Morcuende R, Martinez-Carrasco R. Gas exchange acclimation to elevated CO_2 in upper-sunlit and lower shaded canopy leaves in relation to nitrogen acquisition and partitioning in wheat grown in field chambers. Environmental and Experimental Botany. 2007;**59**:371-380

[3] Samarakoon AB, Gifford RM. Soil water content under plants at high CO_2 concentration and interactions with the direct CO_2 effects: A species comparison. Journal of Biogeography. 1995;**22**:193-202

[4] BassiriRad H, Gutschick VP, Lussenhop J. Root system adjustments: Regulation of plant nutrient uptake and growth responses to elevated CO_2. Oecologia. 2001;**126**:305-320

[5] Norby RJ, Warren JM, Iversen CM, Medlyn BE, McMurtrie RE. CO_2 enhancement of forest productivity constrained by limited nitrogen availability. Proceedings of the National Academy of Sciences. 2010;**107**:19368-19373

[6] Caliman FRB. Enriquecimento com CO_2 por meio de compostagem para a cultura do tomateiro em ambiente protegido. Tese (Doutorado em Fitotecnia) - Programa de Pós-graduação em Fitotecnia, Universidade Federal de Viçosa, Viçosa, MG; 2008

[7] Bennett EM, Carpenter SR, Caraco NF. Human impacts on erodible phosphorus and eutrophication: A global perspective. Bioscience. 2001;**51**:227-234

[8] Peñuelas J, Poulter B, Sardans J, Ciais P, van der Velde M, Bopp L, Boucher O, Godderis Y, Hinsinger P, Llusia J, Nardin E, Vicca S, Obersteiner M, Janssens IA. Human-induced

nitrogen-phosphorus imbalances alter natural and managed ecosystems across the globe. Nature Communications. 2013;**4**:293. DOI: 10.1038/ncomms3934

[9] Bracken MES, Hillebrand H, Borer ET, Seabloom EW, Cebrian J, Cleland EE, Elser JJ, Gruber DS, Harpole WS, Ngai JT, Smith JE. Signature of nutrient limitation and co-limitation: Response of autotroph internal nutrient concentrations to nitrogen and phosphorus additions. Oikos. 2014;**124**:113-121

[10] Yuan ZY, Chen HYH. Negative effects of fertilization on plant nutrient resorption. Ecology. 2015;**96** 373-380

[11] MacDonald GK, Bennett EM, Potter PA, Ramankutty N. Agronomic phosphorus imbalances across the world's croplands. Proceedings of the National Academy of Sciences. 2011;**108**:3086-3091

[12] Peñuelas J, Sardans J, Rivas-ubach A, Janssens IA. The human-induced imbalance between C, N and P in Earth's life system. Global Change Biology. 2012;**18**:3-6

[13] Sardans J, Rivas-Ubach A, Peñuelas J. The C:N:P stoichiometry of organisms and ecosystems in a changing world: A review and perspectives. Perspectives in Plant Ecology. 2012;**14**:33-47

[14] Sterner RW, Elser JJ. Ecological Stoichiometry: The Biology of Elements from Molecules to the Biosphere. Princeton, NJ, USA: Princeton University Press; 2002

[15] Bian J, Berninger JP, Fulton BA, Brooks BW. Nutrient stoichiometry and concentrations influence silver toxicity in the aquatic macrophyte *Lemna gibba*. Science of the Total Environment. 2013;**449**:229-236

[16] Zechmeister-Boltenstern S, Keiblinger KM, Mooshammer M, Peñuelas J, Richter A, Sardans J, Wanek W. The application of ecological stoichiometry to plant-microbial-soil organic matter transformations. Ecological Monographs. 2015;**85**:133-155

[17] Hessen DO, Elser JJ, Sterner RW, Urabe J. Ecological stoichiometry: An elementary approach using basic principles. Limnology and Oceanography. 2013;**58**:2219-2236

[18] Elser JJ, Bracken MES, Cleland EE, Gruner DS, Harpole WS, Hillebrand H, Ngai JT, Seabloom EW, Shurin JB, Smith JE. Global analysis of nitrogen and phosphorus limitation of primary production in freshwater, marine, and terrestrial ecosystems. Ecology Letters. 2007;**10**:1135-1142

[19] Raven JA, Andrews M, Quigg A. The evolution of oligotrophy: Implications for the breeding of crop plants for low input agricultural systems. Annals of Applied Biology. 2005;**146**:261-280

[20] Zhang LX, Bai YF, Han XG. Differential responses of N:P stoichiometry of *Leymuschinensis* and *CarexKorshinskyi* to N additions in a steppe ecosystem in Nei Mongol. Acta Botanica Sinica. 2004;**46**:259-270

[21] Güsewell S. N:P ratios in terrestrial plants: Variation and functional significance. New Phytologist. 2004;**164**:243-266

[22] Koerselman W, Meuleman AFM. The vegetation N:P ratio: A new tool to detect the nature of nutrient limitation. Journal of Applied Ecology. 1996;**33**:1441-1450. DOI: 10.2307/2404783

[23] McGroddy ME, Daufresne T, Hedin LO. Scaling of C:N:P stoichiometry in forests worldwide: Implications of terrestrial Redfield-type ratios. Ecology. 2004;**85**:2390-2401

[24] Kerkhoff AJ, Enquist BJ, Elser JJ, Fagan WF. Plant allometry, stoichiometry and the temperature dependence of primary productivity. Global Ecology and Biogeography. 2005;**14**:585-598

[25] Reich PB, Oleksyn J. Global patterns of plant leaf N and P in relation to temperature and latitude. Proceedings of the National Academy of Sciences. 2004;**101**:11001-11006.

[26] Valladares F, Martínez-Ferri E, Balaguer L, Pérez-Corona E, Manrique E. Low leaf-level response to light and nutrients in Mediterranean evergreen oaks: A conservative resource-use strategy? New Phytologist. 2000;**148**:79-91

[27] Wu Z, Dijkstra P, Koch GW, Peñuelas J, Hungate BA. Responses of terrestrial ecosystems to temperature and precipitation change: A metaanalysis of experimental manipulation. Global Change Biology. 2011;**17**:927-942

[28] Jentsch A, Kreyling J, Elmer M, Gellesch E, Glaser B, Grant K, Hein R, Lara M, Mirzae H, Nadler SE, Nagy L, Otieno D, Pritsch K, Rascher U, Schadler M, Schloter M, Singh BK, Stadler J, Walter J, Wellstein C, Wollecke J, Beierkuhnlein C. Climate extremes initiate ecosystem-regulating functions while maintaining productivity. Journal of Applied Ecology. 2011;**99**:689-702. DOI: 10.1111/j.1365-2745.2011.01817.x

[29] Lloret F, Escudero A, Iriondo JM, Martínez-Vilalta J, Valladares F. Extreme climatic events and vegetation: The role of stabilizing processes. Global Change Biology. 2012;**18**: 797-805

[30] Sardans J, Peñuelas J, Estiarte M, Prieto P. Warming and drought alter C and N concentration, allocation and accumulation in a Mediterranean shrubland. Global Change Biology. 2008b;**14**:2304-2316

[31] González AL, Kominoski JS, Danger M, Ishida S, Iwai N, Rubach A. Can ecological stoichiometry help explain patterns of biological invasion? Oikos. 2010;**119**:779-790

[32] Sardans J, Peñuelas J, Estiarte M. Changes in soil enzymes related to C and N cycle and in soil C and N content under prolonged warming and drought in a Mediterranean shrubland. Applied Soil Ecology. 2008a;**39**:223-235

[33] Matías L, Castro J, Zamora R. Soil-nutrient availability under a global-change scenario in a Mediterranean mountain ecosystem. Global Change Biology. 2011;**17**:1646-1657

[34] Sardans J, Peñuelas J, Estiarte M. Warming and drought alter soil phosphatase activity and soil P availability in a Mediterranean shrubland. Plant and Soil. 2006;**289**:227-238

[35] Lu YW, Duan BL, Zhang XL, Korpelainen H, Berninger F, Li CY. Intraspecific variation in drought response of *Populus cathayana* grown under ambient and enhanced UV-B radiation. Annals of forest Science. 2009;**66**:1-12

[36] Sardans J, Peñuelas J. Drought changes phosphorus and potassium accumulation patterns in an evergreen Mediterranean forest. Functional Ecology. 2007;**21**:191-201

[37] Ferreira V, Goncalves AL, Godbold DL, Canhoto C. Effect of increased atmospheric CO_2 on the performance of an aquatic detritivore through changes in water temperature and litter quality. Global Change Biology. 2010;**16**:3284-3296

[38] Atkin OK, Bruhn D, Hurry VM, Tjoelker MG. The hot and the cold: Unravelling the variable response of plant respiration to temperature. Functional Plant Biology. 2005;**32**: 87-105

[39] Aerts R, Callaghan TV, Dorrepaal E, van Logtestijn RSP, Cornelissen JHC. Seasonal climate manipulations result in species-specific changes in leaf nutrient levels and isotopic composition in a sub-arctic bog. Functional Ecology. 2009;**23**:680-688

[40] Larsen KS, Andresen LC, Beier C, Jonasson S, Albert KR, Ambus P, Arndal MF, Carter M, Christensen S, HolmstruP M, Ibrom A, Kongstad J, Linden LV, Maraldo K, Michelsen A, Mikkelsen TN, Pilegaard, K, Prieme A, Ro-Poulsen H, Schmidt IK, Selsted MB, Stevnbak K. Reduced N cycling in response to elevated CO_2, warming, and drought in a Danish heathland: Synthesizing results of the climate project after two years of treatments. Global Change Biology. 2011;**17**:1884-1899. DOI: 10.1111/j.1365-2486.2010.02351.x

[41] Day TA, Ruhland CT, Xiong FS. Warming increases aboveground plant biomass and C stocks in vascular-plant-dominated Antarctic tundra. Global Change Biology. 2008;**14**: 1827-1843

[42] Ågren GI. The C:N:P stoichiometry of autotrophs–theory and observations. Ecology Letters. 2004;**7**:185-191

[43] Matzek V, Vitousek PM. N:P stoichiometry and protein: RNA ratios in vascular plants: An evaluation of the growth-rate hypothesis. Ecology Letters. 2009;**12**:765-771

[44] Elser JJ, Acharya K, Kyle M, Cotner J, Makino W, Markow T, Watts T, Hobbie S, Fagan W, Schade J, Hood J, Sterner RW. Growth rate–stoichiometry couplings in diverse biota. Ecology Letters. 2003;**6**:936-943. DOI: 10.1046/j.1461-0248.2003.00518.x

[45] Reich PB, Tjoelker MG, Pregitzer KS, Wright IJ, Oleksyn J, Machado JL. Scaling of respiration to nitrogen in leaves, stems and roots of higher land plants. Ecology Letters. 2008;**11**:793-801

[46] Kerkhoff AJ, Enquist BJ. Ecosystem allometry: The scaling of nutrient stocks and primary productivity across plant communities. Ecology Letters. 2006;**9**:419-427

[47] Novotny AM, Schade JD, Hobbie SE, Kay AD, Kyle M, Reich PB, Elser JJ. Stoichiometric response of nitrogen-fixing and non-fixing dicots to manipulations of CO_2, nitrogen, and diversity. Oecologia. 2007;**151**:687-696

[48] Billings SA, Lichter J, Ziegler SE, Hungate BA, Richter DD. A call to investigate drivers of soil organic matter retention vs. mineralization in a high CO_2 world. Soil Biology & Biochemistry. 2010;**42**:665-668

[49] Elser JJ, Dobberfuhl DR, MacKay NA, Schampel JH. Organism size, life history, and N:P stoichiometry. Bioscience. 1996;**46**:674-684

[50] Ågren GI, Weih M. Plant stoichiometry at different scales: Element concentration patterns reflect environment more than genotype. New Phytologist. 1996;**194**:944-952

[51] Elser JJ, Fagan WF, Kerkhoff AJ, Swenson NG, Enquist BJ. Biological stoichiometry of plant production: Metabolism, scaling and ecological response to global change. New Phytologist. 2010;**186**:593-608

[52] Enquist BJ, Kerkhoff AJ, Huxman TE, Economo EP. Adaptive differences in plant physiology and ecosystem paradoxes: Insights from metabolic scaling theory. Global Change Biology. 2007;**13**:591-609

[53] Beardall J, Allen D, Bragg J, Finkel ZV, Flynn KJ, Quigg A, Rees TA, Richardson A, Raven JA. Allometry and stoichiometry of unicellular, colonial and multicellular phytoplankton. New Phytologist. 2009;**181**:295-309

[54] Yuan ZY, Chen HYH, Reich PB. Global-scale latitudinal patterns of plant fine-root nitrogen and phosphorus. Nature Communications. 2011;**2**:344

[55] Zhang H, Wu H, Yu Q, Wang Z, Wei C, Long M, Kattge J, Smith M, Han X. Sampling date, leaf age and root size: Implications for the study of plant C:N:P stoichiometry. Plos One. 2013;**8**:e60360

[56] McJannet C, Keddy P, Pick F. Nitrogen and phosphorus tissue concentrations in 41 wetland plants: A comparison across habitats and functional groups. Functional Ecology. 1995;**9**:231-238

[57] Chapin IIIFS. The mineral nutrition of wild plants. Annual Review of Ecology, Evolution, and Systematics. 1980;**11**:233-260

[58] Alvarez RCF Prado RM, Felisberto G, Deus AF, Oliveira RLL. Soluble silicate and nanosilica applied in oxisol on nutrition of rice plant (Dados não publicados)

[59] Eneji AE, Inanaga S, Muranaka S, Li J, Hattori T, An P, Tsuji W. Growth and nutrient use in four grasses under drought stress as mediated by silicon fertilizers. Journal of Plant Nutrition. 2008;**31**:355-365

[60] Schaller J, Brackhage C, Gessner MO, Bäuker E, Gert dudel E. Silicon supply modifies C:N:P stoichiometry and growth of *Phragmites australis*. Plant Biology. 2012;**14**:392-396

[61] IPCC. Climate Change 2001: The Scientific Basis. New York: Cambridge University Press; 2001

[62] Reich PB, Hobbie SE, Ellsworth DS, West JB, Tilman D, Knops JMH, Naeem S, Trost J. Nitrogen limitation constrains sustainability of ecosystem response to CO_2. Nature. 2006; **440**:922-925

[63] Luo YQ, Hui DF, Zhang DQ. Elevated CO_2 stimulates net accumulations of carbon and nitrogen in land ecosystems: A meta analysis. Ecology. 2006;**87**:53-63

[64] Ferris R, Sabatti M, Miglietta F, Mills RF, Taylor G. Leaf area is stimulated in Populus by free air CO_2 enrichment (POPFACE), through increased cell expansion and production. Plant, Cell & Environment. 2001;**24**:305-315

[65] Hyvonen R, Agren GI, Linder S, Persson T, Cotrufo MF, Ekblad A, Freeman M, Grelle A, Janssens IA, Jarvis PG, Kellomäki S, Lindroth A, Loustau D, Lundmark T, Norby RJ, Oren R, Pilegaard K, Ryan MJ, Sigurdsson BD, Strömgren M, Oijen M, Wallin G. The likely impact of elevated [CO_2], nitrogen deposition, increased temperature and management on carbon sequestration in temperate and boreal forest ecosystems: A literature review. New Phytologist. 2007;**173**:463-480. DOI: 10.1111/j.1469-8137.2007.01967.x

[66] Tcherkez GGB, Farquhar GD, Andrews TJ. Despite slow catalysis and confused substrate specificity, all ribulose bisphosphate carboxylases may be nearly perfectly optimized. Proceedings of the National Academy of Sciences. 2006;**103**:7246-7251

[67] Majeau N, Coleman JR. Effect of CO_2 concentration on carbonic anhydrase and ribulose-1,5-bisphosphate carboxylase/oxygenase expression in pea. Plant Physiology. 1996;**112**:569-574

[68] Ward JK, Kelly JK. Scaling up evolutionary responses to elevated CO_2: Lessons from Arabidopsis. Ecology Letters. 2004;7:427-440

[69] Taub DR, Wang XZ. Why are nitrogen concentrations in plant tissues lower under elevated CO_2? A critical examination of the hypotheses. Journal of Integrative Plant Biology. 2008;**50**:1365-1374

[70] Enquist BJ, Niklas KJ. Global allocation rules for patterns of biomass partitioning in seed plants. Science. 2002;**295**:1517-1520

[71] Vitousek PM, Howarth RW. Nitrogen limitation on land and in the sea: How can it occur? Biogeochemistry. 1991;**13**:87-115

[72] Galloway JN, Levy II H, Kasibhatia PS. Year 2020: Consequences of population growth and development on the deposition of oxidized nitrogen. Ambio. 1994;**23**:120-123

[73] Vitousek PM, Matson PA. Agriculture, the global nitrogen cycle, and trace gas flux. In: Oremland RS, editor. The Biogeochemistry of Global change: Radiative Trace Gases. New York, USA: Chapman & Hall; 1993. pp. 193-208

[74] Falkowski P, Scholes RJ, Boyle E, Canadell J, Canfield D, Elser J, Gruber N, Hibbard K, Hogberg P, Linder S, Mackenzie FT, Moore III B, Pedersen T, Rosenthal Y, Seitzinger S, Smetacek V, Steffen W. The global carbon cycle: A test of our knowledge of Earth as a system. Science. 2000;**290**:291-296. DOI: 10.1126/science.290.5490.291

[75] Kulmatiski A, Vogt KA, Vogt DJ, Wargo PM, Tilley JP, Siccama TG, Sigurdardottir R, Ludwig D. Nitrogen and calcium additions increase forest growth in northeastern USA spruce–fir forests. Canadian Journal of Forest Research. 2007;**37**:1574-1585

Chapter 9

Biogeographical Areas of Hispaniola (Dominican Republic, Republic of Haiti)

Ana Cano Ortiz, Carmelo María Musarella and Eusebio Cano Carmona

Additional information is available at the end of the chapter

http://dx.doi.org/10.5772/intechopen.69081

Abstract

The island of Hispaniola is located between parallels 17 and 19 N and forms part of the Greater Antilles group in the Caribbean region. It covers an area of 76,484 km^2 and has the highest altitudes in the whole Caribbean region. The island consists of two countries: the Dominican Republic and the Republic of Haiti. The flora of both countries has been studied in depth by Liogier and several authors from the Dr. Rafael Ma. Moscoso National Botanical Garden in Santo Domingo; this has enabled us to examine the distribution of 1582 endemic species in 19 areas and several important endemic habitats for conservation: *Lepotogono buchii-Leptochloopsietum virgatae; Crotono astrophori-Leptochloopsietum virgatae; Melocacto pedenalensi-Leptochloopsietum virgatae* and *Solano microphylli-Leptochloopsietum virgatae* pine forests on serpentine belonging to the association *Leptogono buchii-Pinetum occidentalis* and high-mountain pine forests: *Dendropemom phycnophylli-Pinetum occidentalis* and *Cocotrino scopari-Pinetum occidentalis*. Some dry forest communities are of interest, including *Chrysophyllo oliviformi-Sideroxyletum salicifolii* and *Zamio debilis-Metopietum toxiferi*. Based on the floristic analysis and the vegetation study, a biogeographical typology for the island, in which we propose 19 biogeographical areas (BA) has been established.

Keywords: Caribbean, Hispaniola, biogeography, territory, area, flora and vegetation

1. Introduction

The geological history underlying the formation of the island of Hispaniola [1], the great differences in altitude and the wide range of substrates, have all led to the existence of 2050 endemic species distributed across a wide variability of habitats with an endemic nature that must be highlighted for the purposes of conservation. Therefore, in previous works, we started our study with a biogeographic proposal for the Spanish island [2].

We therefore begin our study for a biogeographical proposal for the island of Hispaniola with the general description by Takhtajan [3] on floristic regions in the world, and the work of Rivas Martínez et al. [4] on North and Central America, which establishes the rank of sector for the island of Hispaniola, and includes it within the Neotropical-Austro-American kingdom, the Caribbean-Mesoamerican region and the province of the Antilles. The studies by Borhidi [7] on the island of Cuba [5, 6] all recognise clear differences between the Greater Antilles (Cuba, Hispaniola, Jamaica and Puerto Rico) and the Lesser Antilles, based—among other considerations—fundamentally on the high biodiversity and distribution of species in the *Orchidaceae* family. Together with the existing local studies on the island of Hispaniola [8, 9] and our own recent fieldwork, we can establish a biogeographical typology based on the elemental biogeographical unit or tesela, defined as a variable area—either continuous or discontinuous—with a homogeneous geomorphological and ecological character giving rise to a single type of potential vegetation.

The island of Hispaniola with a territorial extension of 76,486 km^2 is divided between Doninican Republic and Republic of Haiti, it is an island very much studied from the floristic point of view, with very few studies of vegetation, with few studies like the one by Borhidi [7].

The importance of the study is due to the high diversity of endemic species and habitats that are of interest for conservation, with species and habitats subject to high human pressure, despite being a hot spot for the Caribbean.

In general, the island is dominated by areas very antropizadas, especially Haiti, where the human pressure is excessive; while in the Dominican Republic, anthropogenic action is somewhat lower, with areas dedicated to livestock and agriculture. However, there are two large well preserved landscape units: the rainforests of the mountains and the dry sub-deciduous forests.

2. Methodology

We conduct a botanical study from a biogeographical approach; as notwithstanding the numerous botanical and floristic works of investigation by researchers such as Urban, Ekman, Cicero, Donald Dungan, Marcano Fondeur, Jürgen Hoppe, Liogier, Zanoni, Hager, May, Borhidi, Megía, Jiménez, R. García, A. Veloz and others, very few studies have been undertaken from the perspective of phytogeography and vegetation science. Recent works have taken a floristic and physiognomic rather than a phytosociological approach [4, 8–18, 20, 21]; studies using a phytosociological methodology include Refs. [22–31]. The authorship of the species is mentioned only once in the text and is taken from Ref. [36].

The different types of habitats present on the island of Hispaniola are studied. For this purpose, we have performed over 300 years of phytosociological sampling as per Ref. [32]. At the same time, we do a floristic study on the distribution of 1500 endemisms. In order to explain the distribution of the different plant communities, a bibliographical and field-based study was carried out on existing geological materials, and a bioclimatic study of ombrotipos and thermotypes is presented in refs. [4, 33–35].

3. Results

The island's geological origin, the bioclimatic analysis with thermotypes ranging from the infratropical to the supratropical, with semiarid to hyperhumid ombrotypes, the origin of the flora as a result of migratory routes and the past isolation of the various sierras and mountains, all account for the large number of endemic species and habitats. The island has 1284 genera, of which 31 are endemic to Hispaniola: *Zombia, Leptogonum, Arcoa, Neobuchia, Fuertesia, Sarcopilea, Salcedoa, Eupatorina, Vegaea, Coeloneurum, Theophrasta, Haitia, Stevensia, Samuelssonia, Hottea, Tortuella* and *Anacaona*, among others. Several of the endemic genera are monotypes and have a restricted area, such as *Vegaea pungens* Urb., *Zephyranthes ciceroana* M. Mejía & R. García, *Gautheria domingensis* Urb., *M. domingensis, Omphalea ekmanii* Alain, *Gonocalyx tetrapterus* A. Liogier, *Goetzea ekmanii, Reinhardtia paiewonskiana* R.W. Read, T. Zanoni & M. Mejía, *Pseudophoenix ekmanii* Burret and *Salcedoa mirabaliarum* F. Jiménez & L. Katinas; or else local endemic species such as *Pinguicula casabitoana, Fuertesia domingensis* Urb., *Pereskia quisqueyana, M. jimenezii* Alain and *Salvia montecristina.*

There are a total of 5800 species according to [36], a figure that was subsequently extended by [37] to 6000 vascular species distributed in 1284 genera, with an estimated 2050 endemic species. Our study characterises the various biogeographical territories based on 1582 endemic species distributed in 19 areas (A1–A19), together with their own vegetation catenas, which we in turn break down into two subprovinces: the Central subprovince and the Caribbean-Atlantic subprovince, both clearly separate due to differences in their geological, bioclimatic, floristic and vegetation origin.

The province is characterised by a large number of endemic species, of which 114 belong to the family *Melastomataceae* [24]. The presence of a high number of endemic species with a widespread distribution on the island causes the application of Pearson's index to result in a low relation between areas A12 and A16 (r = 1.25), due to their different geological and floristic nature; between A16 and A13 (r = 1.17); and between A12 and A13 (r = 1.23). In this last case, the low relation between the two areas derives from the difference in the number of endemics. Although both zones have calcareous substrates, A13 has suffered greater human impact. A16 and A17 are highly separated, which is unsurprising as the Massif du Nord (A17) is a prolongation of the Central Cordillera range (A16). The frequent presence of calcareous islands and the intense human pressure in A17 is a further reason for their separation, with A17 acquiring a greater similarity with A15 (northwest Haiti).

The Jaccard analysis reveals that areas A12 and A16 have a distance of 0.9, representing a coincidence of only 10% and differences of 90%; this is also the case between A16 and A17, as this analysis corroborates that A17 has greater similarity with A15. The results of Pearson's analysis for areas A12 and A13 are similar to those obtained with the Jaccard analysis [24].

4. Analysis of biogeographical territories (sectors) and areas (districts) (BT, BA)

The Central Cordillera is characterised by a predominance of siliceous materials and a tropical rainy and tropical pluviseasonal macrobioclimate on the summits, occasionally tropical xeric at the base. All this has led to the development of a particular endemic flora with 451 different endemic species and vegetation units [11]. This biogeographical territory (BT) contains a single area, A16, Central-Eastern, occupying the Central Cordillera (Dominican Republic), and dominated by siliceous materials with slight inclusions of serpentines in the easternmost area, representing the transition to the Yamasense biogeographical unit. The thermotype ranges between the infra and supratropical, and the dominant ombrotype is humid-hyperhumid. The penetration of the trade winds causes the presence of both broadleaf rainforest and cloud forest towards the mid-mountain, with a dominance of species from the genera *Prestoea, Magnolia, Didymopanax, Cyathea*; while the high-mountain areas beyond the reach of the trade winds are home to *Pinus occidentalis*, a forest belonging to the endemic habitat *Dendropemon phycnophylli–Pinetum occidentalis* Cano, Veloz & Cano-Ortiz 2011, alternating with hemicryptophytic communities of *Danthonia domingensis* **Figure 1**.

In previous works, we proposed the following biogeographical territories (BT) (biogeographical sectors) for the Caribbean-Atlantic subprovince: 2.1. Bahoruco-Hottense (A12, A13); 2.2. Neiba-Matheux-North-western (A14, A15, A17 and A19); 2.3. Azua- San Juán-.Hoya

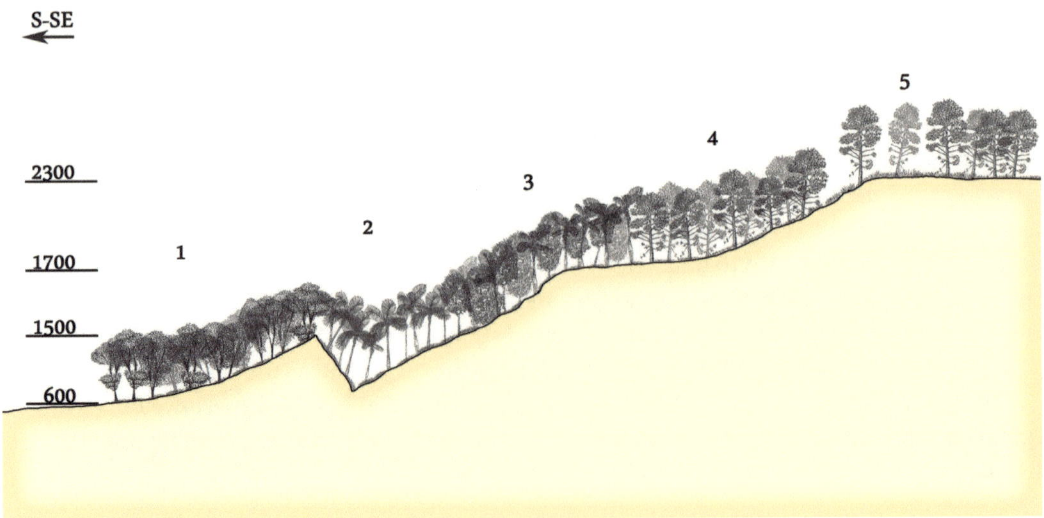

Figure 1. Vegetation catena of the Central Cordillera 1. Subhumid broadleaf forest. 2. Sierran palm forest. Community of *Prestoea montana* and *Cyateas* (palms). 3. Community of 'palo de viento' *Didymopanax tremulus* and isolated individuals of *Prestoea montana*. 4. *Pinus occidentalis* pine forest belonging to the association *Dendropemon phycnophylli–Pinetum occidentalis*. 5. Hemicryptophytic high-mountain grassland of *D. domingensis* among the cleared pine forest of *Pinus occidentalis*.

Enriquillo-Port au Prince-Artiobonite-Gonaivës (A9, A10, A11 and A18); 2.4. Caribbean-Cibense (A3, A7 and A8); 2.5. Northern (A1, A2, A4, A5 and A6) [24].

BT-2. 1. The Bahoruco-Hottense district includes two areas or districts (A12 and A13). The Sierra de Bahoruco and its continuation in the Massif de la Selle and de la Hotte in Haiti have a similar geological origin and frequently suffer the impact of Caribbean hurricanes. The ombrotype in these territories ranges from subhumid to hyperhumid, leading to a predominance of broadleaf cloud forest, sierran palm forests of *Prestoea montana*, cloud forest of *Magnolia* and *Didymopanax*, and—in the supratropical thermotype on the summits—a pine forest of *P. occidentalis*, belonging to the association *Coccotrino Scopari-Pinetum occidentalis* Cano, Veloz & Cano-Ortiz 2011. The general vegetation catena characterising this biogeographical territory is therefore conditioned whether it is a dry forest, broadleaf forest, cloud forest or high-mountain pine forest. In addition, due to its high rate of endemic species, this biogeographical territory is of interest for conservation. We are unaware of the existence of *Podocarpus aristulatus* Parl. and *Ocotea wrightii* (Meisn.) Mez in the Bahoruco-Hottense sector; and this BT thus reveals significant differences when compared with the Neiba-Matheux-North-western sector. The relation between the two areas (A12 and A13) in this BT is low, as they present a certain number of different endemic species with an $r = 1.23$ **Figure 2.**

BA-A12. The Bahoruco-La Selle district occupies calcareous mountain ranges with an occasionally supratropical thermotype. There is a broadleaf cloud forest of *M. hamorii* and *D. tremulus,*

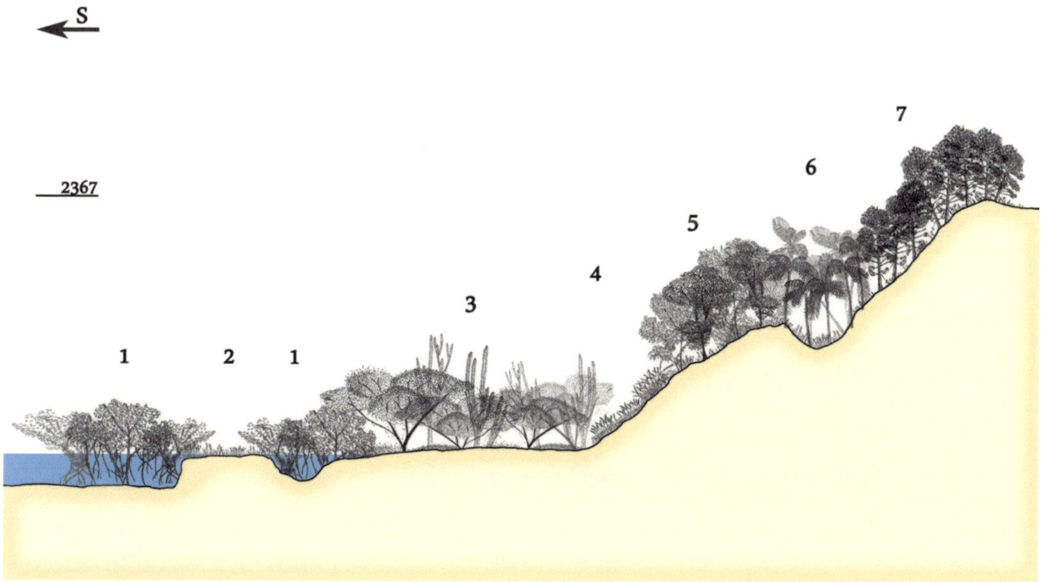

Figure 2. Vegetation catena in the Sierra de Bahoruco. 1. Mangrove forests of *Machaerio lunati-Rhizophoretum manglis*. 2. Salt marsh communities in the class *Batidi-Salicornietea*. 3. Dry forest of *Pilosocereus polygonus* and *Acacia sckleroxyla*. 4. Hemicryptophytic grassland of *Melocacto pedernalensis-Leptochloopsietum virgatae*. 5. Cloud forest of *Magnolia hamorii* and *Didymopanax tremulus*. 6. Sierran palm forest (forest of *Prestoea montana* and *Cyathea arborea*). 7. Pine forest of *Coccotrino scopari-Pinetum occidentalis*.

while on rainier sites and in gorges, there is a presence of formations of *P. montana*. This unit is home to forests of *M. hamorii* growing between 950 and 1500 m, as precipitation exceeds 2000 mm. These forests are characterised by *M. hamorii, L. bahorucanus, Mikania venosa* Alain, *C. domingensis, Rondeletia ochracea* Urb., *P. guadalupensis, H. domingensis, Arthrostylidium sarmentosum* Pilger, *Weinmannia pinnata* L., *M. ovatum, Vriesea tuercheimii, D. tremulus, Meriania involucrata* (Desr.) Naud. and *Polygala fuertesii* (Urb.) Blake. The same cloud forest at higher altitudes, where the pressure of the wind is greater, is enriched by *D. tremulus*, and on rainier sites and in moist gorges by the sierran palm forest of *P. montana*. A pine forest of *P. occidentalis* is found growing in the supratropical thermotype, with *C. scoparia, A. intermixta, N. domingensis, Eupatorium sinuatum* Lam. *var viscigerum* Urb. & Ekm., *Staurogyne repens, G. ruolphiodes var haitiensis* and *S. barahonenis*, belonging to the association *Cocotrino scopari–Pinetum occidentalis* Cano, Veloz & Cano-Ortiz 2011.

In basal zones such as the Procurrente de Barahona, Ceitillan and Pedernales, the ombrotype is semiarid and the thermotype is infratropical. There is a predominantly dry forest, with a floristic composition comprising *Lomandra hystrix, P. polygonus, Ceratocystis moniliformis, Antillesoma antillarum, Coriandria caribaea* and *Melocactus pedernalensis*, in whose clearings there is a hemicryptophytic and endemic community of *Melocaptoa pedernalensis-Leptochloopsietum virgatae* Cano, et al. [24, 25]. In coastal areas, it is worth highlighting the presence of mangrove forests of *R. mangle, L. racemosa* and *A. germinans*, enriched towards drier areas with *C. erectus*. In these territories, the mangrove forest alternates with halophilous communities of *S. portulacastrum*. This area has a high rate of endemic species, with 693 endemic plants.

BA-A13. The Hottense district is characterised by calcareous substrates whose geological origin is similar to that of La Selle and Bahoruco. It is located at the end of the southwest peninsula (Haiti), and has 171 endemic plants, but in lesser numbers than in the Bahoruco-La Selle area. However, this biogeographical unit is home to the endemic genus *Hottea*, which is distributed throughout the biogeographical units A12, A13 and A14; the highest numbers of endemic species in this genus are found in A13. This biogeographical unit has a thermotype that ranges between the infra and mesotropical, and the ombrotype is dry in the basal areas to humid on the summits of La Hotte.

BT-2.2. The Neiba-Matheux-North-western sector has four districts (A14, A15, A17 and A19). It is floristically characterised by the presence of 90 endemic species such as *Guettarda oxyphylla* Urb. and *Chionanthus dictyophyllus* (Urb.) Stearn, with its own vegetation catenas ranging from the dry to the subhumid and cloudy, with pine forests of *P. occidentalis* on the summits. Two of the areas in this biogeographical territory (A14 and A15) have a relation r = 0.93, indicating major floristic differences between both biogeographical units **Figure 3**.

BA-A14. The Neiba-Matheux district covers the calcareous mountain ranges of Neiba, Matheux and Noires with an altitude of 1793 m and a dry, subhumid-humid ombrotype and an infra, thermo and mesotropical thermotype, which is occasionally supratropical in the Massif des Montagnes Noires. It is home to a rare broadleaf forest of *P. aristulatus*, a cloud forest of *P. montana* and a forest of *D. tremulus* which is enriched with *P. aristulatus, O. wrightii, Persea krugii* Mez and *Brunellia comocladifolia* H. & B. In the highest sites in the Neiba range, it is still possible to find pine forests of *P. ocidentalis* on calcareous substrates over a limited

Figure 3. Vegetation catena in the Sierra de Neiba. 1. Forest of *Pinus occidentalis*. 2. Sierran palm forest of *Didymopanax tremulus* and *Podocarpus aristulatus*; 3. Broadleaf mahogany forest of *Swietenia mahagoni*.

extension. This area reveals a certain influence of the Central province due to the presence of *P. aristulatus*, which is also located near Valle Nuevo (Central Cordillera) and the absence of *M. hamorii*; the presence of the pine forest of *P. occidentalis* connects it with Bahoruco. There are 27 endemic species exclusive to this area.

BA-A15. This district is located in the northwest of the Republic of Haiti and also has a predominance of carbonated rocks. These territories are exposed to the trade winds from the Atlantic. However, as they contain no major elevations—the maximum altitude of around 840 m—the dominant ombrotype is subhumid. West-facing areas connecting with Gonaive and Hene Bay have a dry ombrotype, as this is an area of shade, and a thermotype ranging between infra and mesotropical. The floristic character is based on the presence of 59 endemic species.

BA-A17. The Central-Western district occupies the whole of the Massif du Nord in Haiti. This mountain is a prolongation of the Central Cordillera. In this case, there is also a dominance of siliceous materials with the inclusion of basic substrates; the thermotype is infra to mesotropical, and the ombrotype ranges from subhumid to humid. We collected fewer endemic species (60) in this biogeographical unit than in the Central-Eastern unit (A16), as these areas are highly altered, as is generally the case in the whole of the Republic of Haiti, which has suffered widespread deforestation throughout its history.

BA-A19. The Tortuga Island district is calcareous in nature and located in the north of Haiti, at a maximum altitude of 378 m so the trade winds only reach the highest areas. The vast majority of the territory has a dry ombrotype, occasionally becoming subhumid-humid. The relation of A19 with the nearest areas—A3 and A15—is $r = 1.02$ and $r = 0.89$. In spite of its small size, the presence of the monotypical genus *Tortuella abietifolia* Urb. & Ekman and 15 exclusive endemic species justifies its consideration as a biogeographical unit in itself.

BT-2.3. The Azua-San Juán-Hoya Enriquillo-Port au Prince-Artiobonite-Gonaivës sector (A9, A10, A11 and A18) covers all the low-lying areas in the south of the Dominican Republic and west of Haiti. These territories are differentiated from the Procurrente de Barahona as they have soft deposit materials, despite their similar infra and thermotropical thermotype and semiarid-dry ombrotype. However, in this case, there is an occasional presence of the subhumid ombrotype on the heights of the Sierra Martín García, which represents a small island surrounded by dry forest, distinguishing this area from the previous one. Although most of the territory is dominated by dry forest, there is an occasional presence of broadleaf forest on the summits of Martín García. Unlike in Bahoruco and Neiba, there are no formations of *P. occidentalis*. The dry forest in this biogeographical unit is dominated by the species *P. polygonus, L. hystrix, A. antillarum, Mimosa diplotricha* C. Wright ex Sauvalle, *Brya buxifolia* (Murr.) Urb., *N. paniculada, Thouinia domingensis* Urb. & Radlk., *Solanum microphyllum* (Lam.) G. Don., *Coccotrinax spissa* Bailey, *A. skleroxyla, Scolosanthus triacanthus* (Spreng.) DC., *C. moniliformis, M. lemairei* and *C. caribaea*. In view of the differences in flora, vegetation, geology, ombrotype and thermotype, these areas should be treated as specific biogeographical territories. In all cases, the relations between the four areas proposed have a value of *r* in Pearson's index of equal to or less than 1, as they share very few endemic species.

BA-A9. The Azua-Sán Juán-Hoya Herniquillo district is an area that extends from Bani and Azua towards the Sán Juan river valley as far as the border with Haiti, where it becomes what is known as the Central Plain in Haiti, with higher elevations. The Cordillera Central in the Sierra de Neiba is separated through the Sán Juán valley. These semiarid-dry territories border the Sierra de Neiba along the south, and extend along Lake Herniquillo to Jimani and Malpaso. From a geological point of view, there is a predominance of soft materials of a Quaternary nature with gypsum islands in the areas near Herniquillo Lake. There is a constant infra and thermotropical thermotype and a semiarid-dry ombrotype. This district has 85 endemic species, some as emblematic as *N. paniculada, M. lemairei, Acacia barahonesis* Urb. and *Zanthoxylum azuense* (Urb. & Ekm.) Jiménez, which give the territory its distinctive character. This area is characterised by the presence of habitats such as the association *Solano microphylli-Leptochloopsietum virgatae* Cano, et al. [24, 25] **Figure 4**.

BA-A10. The Central Plain district (Haiti) ascends through the area of the Sán Juán valley, past the upper stretch of the Artibonite River at altitudes of 100 m. There exists a territory with no natural vegetation that is used for agriculture (Haiti). This is the location of the Central Plain in Haiti, standing at a height of 300–400 m and separating the Massif des Montagnes Noires at 1793 m and the calcareous substrates in the siliceous Central Cordillera (Massif du Nord),

http://dx.doi.org/10.5772/intechopen.69081

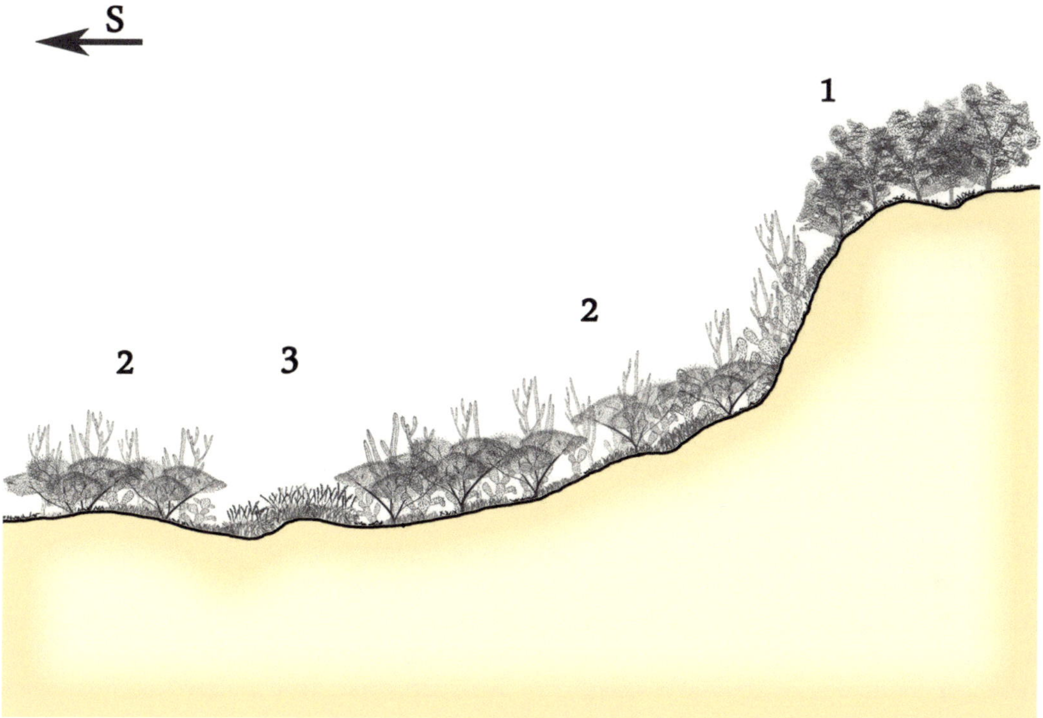

Figure 4. Vegetation catena of the Azua district. 1. Broadleaf forest of *Acacia skleroxyla* and *Coccothrinax boschiana*. 2. Dry forest of *Neoabbottio paniculatae-Guaiacetum officinalis*. 3. Hemicryptophytic community of *Solana microphylli-Leptoch-loopsietum virgatae.*

with maximum altitudes of 1210 m. This district (A10) has calcareous clayey substrates, a thermotropical thermotype and a dry ombrotype, and differs from the semiarid-dry forest in unit A9. This difference is evidenced in the values of Pearson's index—r = 0.91—as the Central Plain in Haiti has higher rainfall, and in its eight endemic species: *Bumelia picardae* Urb., *Carpodiptera hexaptera* Urb. & Ekm, *Dorstenia flagellifera* Urb. & Ekman, *Malpighia aquifolia* L., *Malpighia setosa* Spreng., *Phenax pauciflorus* Urb., *Plumeria paulinae* Urb. and *Thouinidium pinnatum* (Turp.) Radlk.

BA-A11. The Port au Prince-Arbiobonite-Gonaives district is past Jimini (Dominican Republic), approaching the border with Haiti and entering the plain of Port au Prince, where the materials continue to be soft and Quaternary in origin. The thermotype is infra and thermotropical, and the ombrotype is semiarid-dry due to the lack of rain, as these are areas of shadow. The Sierras de Bahoruco, La Selle and Neiba act as a barrier in the south, and those of Matheux, Noires and the Central Cordillera do the same in the northeast. The altitudes range between 0 and 100 m. These are areas with scarcely any natural vegetation as they are used for agriculture; the semiarid-dry character of the territory is prolonged the length of the northern fringe of the Massif de la Selle and de la Hotte, from Port au Prince bay to Gran Caimite, an island that is also part of this biogeographical unit. The territory extends to the southwest of the Massif de Matheux and borders these mountains before entering the Artiobonite river valley as far as the locality of Mirebalais, and northwards towards Gonaives. There are 70 endemic species in these territories, which form part of the different habitats in the dry forest in this biogeographical unit: *Catesbaea sphaerocarpa* Urb., *C. dictyophyllus, Guettarda multinervis* Urb., *Stigmaphyllon haitiense* Urb. & Ndz. and *Psychotria haitiensis* Urb.

BA-A18. Gonave Island is located in the middle of Port au Prince bay, with altitudes of 702 m. It is practically devoid of natural vegetation as a result of intense human pressure. This satellite island of Hispaniola has an infra and thermotropical thermotype and a semiarid-dry ombrotype. Due to its isolation, the floristic analysis reveals the presence of 20 endemic species, of which the following are exclusive to the island: *Mouriri gonavensis* Urb. & Ekman, *Solanum aquartia* Dunal *var luxurians* (O.E. Schulz) Alain, *Dendropemon gonavensis* Urb., *Dendropemon spathulatus* Urb. & Ekman, *Galactia caimitensis* Urb. & Ekman, *Isidorea gonavensis* Aiello & Borhidi and *Pilea dispar* Urb. = (*Pilea gonavensis* Urb.). It shares some endemic species with unit A11—with a Pearson's index of $r = 0.83$—and the same type of vegetation (dry forest). We therefore include it in the same biogeographical sector.

BT-2.4. Caribbean-Cibense (A3, A7 and A8). This biogeographical territory comprises three geomorphological units that include the Eastern Coastal Plain, the Sierras de Yamasá and Prieta, and the Cibao Valley. The eastern coastal areas on the shores of the Caribbean have an altitude of less than 100 m, and are coralline in origin; in contrast, in the Cibao Valley, there is a dominance of alluvial materials, Miocene conglomerates, schists, Miocene loams, limestone hills, clays and calcareous loams. Both geomorphological units are separated by the Sierras of Yamasá and Prieta, in which there is an amalgam of substrates, as this is a crossroads between the Central Cordillera, Los Haitises, the Eastern Coastal Plain and the Cibao Valley. There are frequent serpentines, which can also be found in the province of Dabajón in the Cibao Valley, leading to the widespread presence of a dry spiny forest and a pine forest. The thermotype for this BT ranges from infra to thermotropical and the ombrotype is semiarid to subhumid; however, due to the type of substrate—serpentines and perforated coralline limestones—the territory behaves as dry. All this causes the predominance of a dry forest. This biogeographical unit has particular vegetation associations and catenas. The general vegetation catena is the dry and semi-deciduous forest of *M. toxiferum, S. mahagonii* and *C. diversifolia*. It has a large number of endemic species, distributed in the three biogeographical areas. The relation between the areas in this BT is: $r = 0.95$ for A3–A7, $r = 0.97$ for A3–A8 and $r = 1.01$ for A7–A8 **Figure 5**.

BA-A3. The Cibao Valley district is characterised by a predominance of Miocene limestone, loams and conglomerates, along with the serpentines of Dabajón. The infra and thermotropical thermotype and the semiarid-dry ombrotype have produced a dry forest flora and vegetation, and the presence of 67 endemic plants in this area. The endemic dry forest in the territory comprises a community of *L. hystrix* and *Croton astrophorus* Urb., *C. caribaea, Mammillaria prolifera* (Mill.) Haw., *Phyllostilon brassiliensis* Capan., *H. nashii, P. polygonus, B. buxifolia, Opuntia antillana* Britt. & Rose, *C. moniliformis, Lantana camara* L., *Turnera difusa* Willd. ex Schult, *Abutilon abutiloides* (Jacq.) Garcke, *O. dillenii, Malpighia cnide* K. Spreng., *C. poitaei, S. triacanthus, Erytroxylum rotundifolium* Lunam, *Croton discolor* Willd., *A. antillarum, M. lemairei,*

http://dx.doi.org/10.5772/intechopen.69081

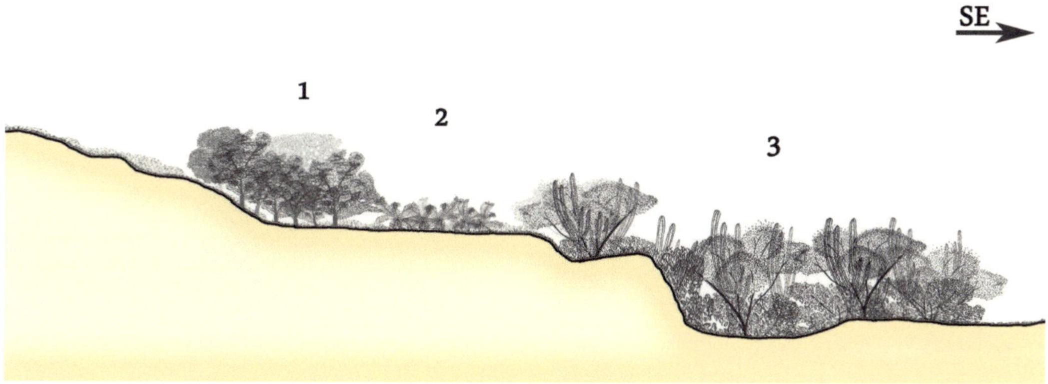

Figure 5. Vegetation catena of the Caribbean coastal unit. 1. Association *Zamio debilis-Metopietum toxiferi*. 2. Community of *Zamia debilis*. 3. Association *Chrysophyllo oliviformi-Sideroxyletum salicifolii*.

Maytenus buxifolia (A. Rich.) Griseb., *L. virgata, Phyllostilon rhamnoides* (J. Poiss.) Taub, *A. tortuosa, Caesalpinia coriaria* (Jacq.) Willd., *Caesalpinia buchii* Urb. and *Anthirea montecristina* Urb. & Ekm. [29]. Along with this dry forest it is frequent to find habitats of *Leptochloopsis virgate* and *Crotono astrophori-Leptochloopsietum virgatae* Cano, et al. [24, 25], and mangrove forests of *Rhabdadenio biflori-Laguncularietum racemosae* Cano et al. [27].

In serpenticolous territories, there is a pine forest or community of *P. occidentales, Calliandra haematomma* (Benth.) Benth., *Tabebuia berterii* (P.DC.) Britt, *Chrysophyllum. oliviforme* L., *Psychotria dolichocalix* Urb. & Ekm., *Smilax habanensis* Jacq., *Sideroxylon cubense* (Griseb.) Penn., *Miconia laevigata* (L.) DC. = (*Miconia pyramidalis* (Desr.) DC., *Croton linearis* Jacq., *Rondeletia cristi* Urb., *O. ilicifolia, Leptogonum buchii* Urb., *Guettarda pungens* Urb., *Ternstroemia peduncularis* A. DC., *Randia aculeata* L., *Byrsonima crassifolia* (L.) HBK. and *L. camara*. This is an exclusive habitat in this biogeographical unit that, along with other communities, serves to establish the differences with the other biogeographical territories: *Leptogono buchi-Pinetum occidentalis* Cano, et al. [26, 19] **Figure 6**.

The Eastern Caribbean area occupies the whole of the coastal plain overlooking the Caribbean Sea. Its coralline origin causes the substrate to be highly porous, and although the rainfall is over 800 mm, the territory acts as dry. There is a strong floristic similarity between these territories in the Cibao Valley and the areas in the southwest, and to a lesser degree between this area and that of Yamasá. There are also differences between the flora, habitats and uses of the territory and the dry areas in the southwest, as the territories on the eastern coastal plain are essentially used for the cultivation of sugarcane and coconut. These eastern plains must therefore be treated as specific biogeographical areas. There are 60 endemic species of flora, some of which have problems of conservation, as is the case of *P. quisqueyana*. In terms of vegetation, it has its own characteristic plant communities depending on the substrate. The dry forest occurs when the soil is thin and porous, but if the soil is deep, these dry phytocoenoses

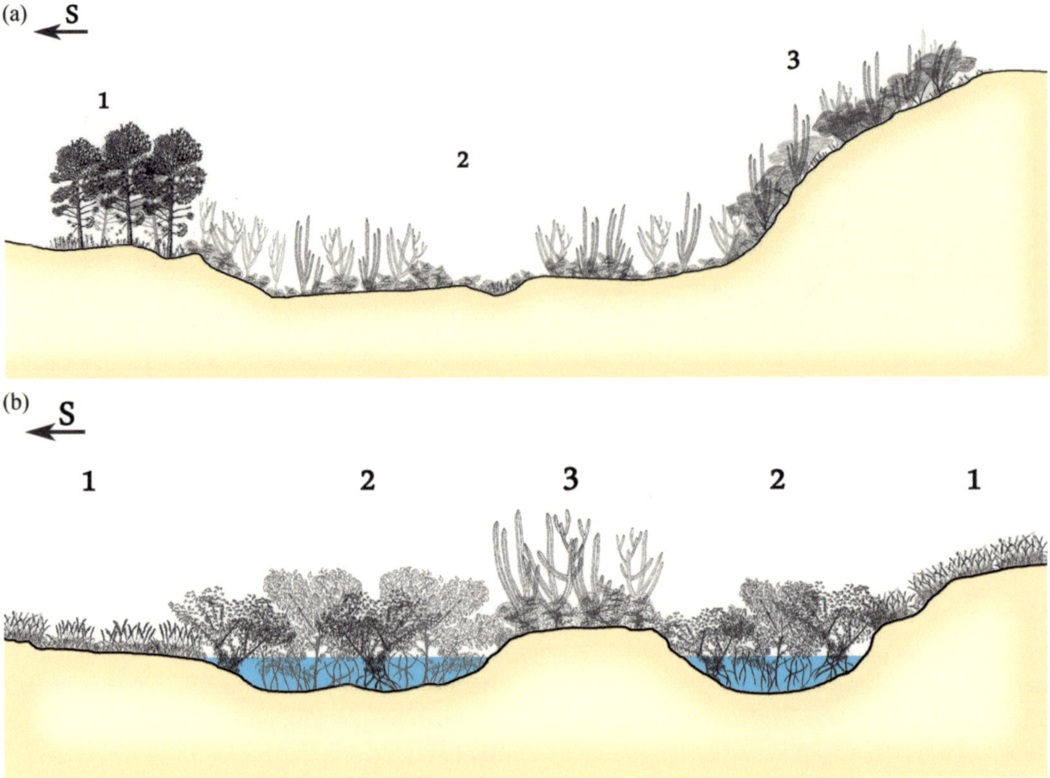

Figure 6. a. Vegetation catena of the Cibao Valley. 1. Association *Leptogono buchii-Pinetum occidentalis*. 2 and 3. Dry forest of *Harrhisio nashii-Prosopidetum juliflorae* [29]. b. Vegetation catena of the Cibao Valley. 1. Association *Crotono astrophori-Leptochloopsietum virgatae*. 2. Association *Machaerio lunati-Rhizophoretum manglis* and *Rhabdadenio biflori-Laguncularietum racemosae*. 3. Dry forest of *Harrhisio nashii-Prosopidetum juliflorae*.

become transformed in semi-deciduous forests in transition between the dry and evergreen forest, comprising a forest of *S. mahagonii, M. toxiferum, Krugiodendron ferreum* (Vahl) Urb., *C. diversifolia, Guaiacum sanctum* L., *Thouinia trifoliata* Poit., *Z. debilis, Coccotrinax barbadensis* (Lodd ex Mart.) Sarg., *Exostema caribaeum* (Jacq.) R. & S., *Sideroxylon salicifolium* (L.) Lam., *C. oliviforme, A. bilobata* and others. On sites with more intense water runoff, there is a dry forest of *Sideroxylon foetidissimum* (Jacq.) Cron., *P. quisqueyana, P. polygonus, L. weingartianus, B. simaruba, Clusia rosea* Jacq., *S. salicifolium, Celtis trinervia* Lam., *B. buceras, Cissus oblongo-lanceolata* (Krug & Urb.) Urb., *Ficus citrifolia* P. Mill., *C. diversifolia, G. sanctum, A. skleroxyla, M. jimenezii, P. unguis-cati, C. oliviforme, K. ferreum, Guapira fragrans* (Dum. Cours.) Little and *Capparis cynophallophora* L. Forests recently diagnosed by us are *Chrysophyllo oliviformi-Sideroxyletum salicifolii* Cano & Veloz 2012 and *Zamio debilis-Metopietum toxiferi* Cano & Veloz 2012. The mangrove forests in the association *Sthalio monospermae-Laguncularietum racemosae* Cano et al. 2012 give this eastern biogeographical area its characteristic appearance.

BA-A8. The Yamasá district is a complex geomorphological unit occupying the Sierra Prieta and Yamasa ranges. It has siliceous, limestone and serpentine substrates, the last of which cause the appearance of endemic serpenticolous communities with a spiny character. This is a xerophytic high shrubland, and throughout the Quaternary era, this territory served as a route for the passage of species between the xeric areas in the Cibao Valley and the Eastern Coastal Plain and the xeric areas in the southwest. The presence of serpenticolous elements and endemic habitats in our study, such as the community of *C. haematomma, Phyllanthus nummularioides* Muell. Arg., *Caliptrogenia biflora* Alain, *Eugenia crenulata* (Sw.) Willd., *Coccotrinax argentea* (Lodd.) Sarg., *L. buchii, Coccoloba nodosa* Lindau, *Coccoloba jimenezii* Alain, *Croton impressus* Urb., *T. peduncularis, Garcinia glaucescens* Alain & M. Mejía, *Scolosanthus densiflorus* Urb., *Rondeletia berteroana* DC., *Oplonia spinosa* (Jacq.) Raf., *Eugenia dictyophylla* Urb., *Pictetia spinifolia* (Desv.) Urban and *Z. debilis*, serves as differentiating elements to establish the Yamasense area. These floristic peculiarities are related to the origin of the territory, which is why different connection forces can be established with the neighbouring territories in the various statistical studies on the rates of endemic species. Although these considerations might suggest a wide biogeographical territory, the absence of its own vegetation catenas, the fact it has a xerophytic vegetation and has served as a migratory route between the biogeographical area of the Eastern Caribbean and the Cibao Valley all lead us to propose this as a highly original biogeographical area **Figure 7**.

BT-2.5. The areas in the north of the island include five biogeographical districts (A1, A2, A4, A5 and A6) comprising the Northern Cordillera, the Samaná Pensinsula and the Eastern Cordillera, the last of which includes Los Haitises. The dominant materials are limestones or coralline rocks, although on the Atlantic coast to the north of the Northern Cordillera, there are islands of serpentines (Puerto Plata and Gaspar Hernández). Although the value of the It/Itc is mitigated by the effects of the trade winds, the thermotype continues to be infra, thermo and mesotropical; the ombrotype in this case ranges between the subhumid in the basal areas and the hyperhumid in territories more exposed to the trade winds. The macrobioclimate is tropical Caribbean-Mesoamerican Pluvial, and there are therefore no dry sites. The spiny forest occurs only in places with serpentines, as the territory acts as dry. The diversity of substrates, the bioclimate and the different dating of the areas accounts for the presence of 154 endemic species.

This territory has a predominance of ombrophilous forest with a rainy character due to the intense influence of the trade winds. This produces a dominance of a broadleaf evergreen forest with well-conserved formations of *P. montana* in Loma Diego de Ocampo, forests of *M. abbottii* to the northeast of the Northern Cordillera and, in somewhat less rainy areas, mahogany forests of *S. mahagoni* and *C. diversifolia*. In addition to the differences in flora and habitats with the rest of the territories, this biogeographical unit lacks the pine forests of *P. occidentalis*, typical of Bahoruco, Neiba and the Central province. In swampy freshwater areas, there are frequent coastal forests of *Pterocarpus officinalis* Jacq., and mangrove forests of *L. racemosa, A. germinans* and *Rhyzophora mangle* L., and to a lesser extent *C. erectus*. In all cases, the relation between the proposed biogeographical units has a Pearson's index of equal to or less than 1, and in some situations, the value of r is very low—$r = 0.73$ for A2–A4 and $r = 0.81$ for A4–A6—indicating a high degree of similarity between the two units **Figure 8**.

BA-A1. The Northern Cordillera district borders the Atlantic Coastal Plain to the north and the Cibao Valley to the south, and is the most recent mountain range on the whole island. There

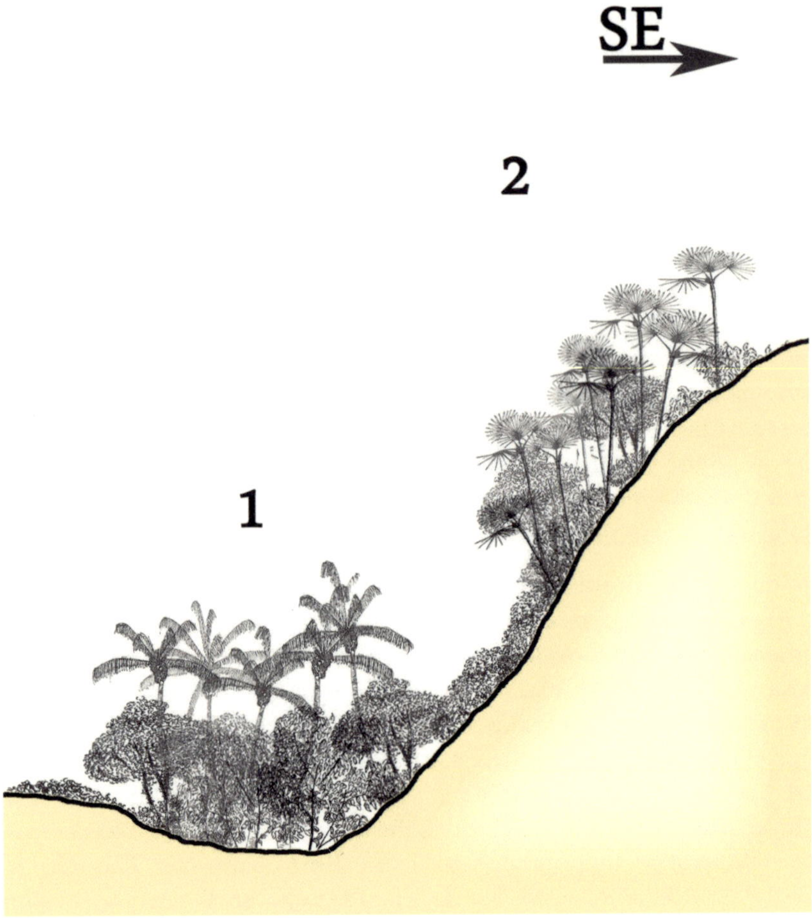

Figure 7. Vegetation catena of the Sierra de Yamasá. 1. Association *Coccotrino argentei-Tabebuietum berterii*. 2. Tangled scrubland of *Garcinio glaucescentis-Phyllanthetum numularioidis* [28].

is a predominance of limestone, schists, and volcanic and metamorphic rocks; the thermotype ranges from infra to mesotropical and the ombrotype is subhumid to hyperhumid. From the floristic point of view, we found 39 endemic species in this unit, representing one of the lowest rates on the island: *Coccotrinax boschiana* M. Mejía & R. García, *Eupatorium trichospermoides* Alain, *Gochnatia microcephala* (Griseb.) Jervis & Alain *var buchii* (Urb.) Alain, *Gonolobus domingensis* Alain, *Justicia spinosissima* Alain, *Sagraea abbottii* (Urb.) Alain, *Cytharexylum alainii* Moldelke, *Mecranium septentrionale* Stean, *Mikania platyloba* Urb. & Ekm. and others. The dominant vegetation is the sierran palm forest of *P. montana, C. racemiflora, D. tremulus, C. clusioides, Cyathea abbottii* Mason, *D. arboresus, T. occidentalis* and *O. capitatus*. On somewhat less cloudy sites and therefore at lower altitudes, there are forests of *M. abbottii*, a species that is also found in the eastern areas of the Central Cordillera and in Sierra Prieta. This species of *M. abbottii* is accompanied by *C. racemiflora, O. leucoxylon* and *S. berteriana*. The epiphyte *Vriesea ringens* (Griseb.) Harms is widespread in these forests, while remnants of mahogany forests

http://dx.doi.org/10.5772/intechopen.69081

Figure 8. Vegetation catena of the Northern Cordillera. 1. Coconut cultivation. 2. Association *Leptogono buchii-Leptochloopsietum virgatae*. 3. Broadleaf mahogany forest of *Swietenia mahagoni*. 4. Sierran palm forest of *Prestoea montana*.

that have been highly altered by humans can be found in the drier areas at the foot of the mountain range. There are only small semi-deciduous copses of *S. mahagonii, C. diversifolia, Zanthoxylum martinicense* (Lam.) DC., *O. leucoxylon, Securidaca virgata* Sw., *Calophyllum calaba* L., *C. argenteum, C. oliviforme* and *G. guidonia*. In this case, the vegetation catena corresponds to a semi-deciduous mahogany forest, followed by a forest of *M. abbottii* and culminating in the more ombrophilous forest of *P. montana*.

BA-A2. The Coastal-Atlantic district is formed by small alluvial valleys of rivers with gentle gradients, with frequent marshes, isolated limestone and reef limestone. It is located to the north of the Northern Cordillera where there is a frequent presence of coconut, coffee and cocoa cultivation in addition to areas of cattle ('potreros' or pastures), so the natural vegetation is highly altered. However, there are 62 endemic species. The thermotype is infratropical and the ombrotype is subhumid and even humid, although the presence of serpentines in Puerto Plata and Gaspar Hernández causes soil xericity. We therefore include the spiny forest in the dry forest, characterised by the presence of specific plant communities such as *Zombia antillarum* (Desc. & Jacks.) Bailey and *S. cubense*, with a frequent presence in this type of forest of *L. buchii, Ouratea ilicifolia* (P.DC.) Bail., *C. sidaefolius, Eugenia maleolens* Pers. = (*Eugenia foetida* Poir.), *Jacquinia umbellata* DC., *C. jimenezii, R. aculeata, M. buxifolia, C. biflora, Vitex heptafila* A. L. Juss., *M. toxiferum, L. virgata, Cordia lima* (Desv.) R. & S., *Tabebuia polyantha* Urb. & Ekm., *C. haematomma, Diospyros caribaea* (A. DC.) Standl., *E. crenulata, C. oliviforme, C. ferrugineum, Bromelia pinguim* L., *Byrsonima spicata* (Cav.) HBK., *Poitaea galegoides* Vent., *Coccoloba pubescens* L., *Eugenia odorata* Berg and *C. linearis*. This is a tall serpentinicolous shrubland (copse) with 60–80% coverage, an average height of 3–4 m and abundant floristic diversity, located in Puerto Plata and Gaspar Hernández in infratropical subhumid-humid areas. This endemic habitat in the Coastal-Atlantic unit belongs to the association *Zombio antillari-Leptogonetum* *buchii* [28]. In the rest of the territory, the potential forest consists of *Swieteania mahagoni* and *C. diversifolia*. An important feature in this area is the presence of the Gran Estero, developed in the last 400–500 years from deposits of materials from the Northern Cordillera. This area is subject to frequent flooding, and is home to a forest of *P. officinalis* belonging to the association *Roystoneo hispaniolanae-Pterocarpetum officinalis* Cano, Veloz, Cano-Ortiz & Esteban 2009. It represents the outer edge of the mangrove forests of *R. mangle* that are typical in the broad channels and in Samaná Bay [23].

BA-A4. The Samaná Peninsula was isolated from the rest of the territory until 300–400 years ago. It constitutes a geomorphological unit dominated by karstic and limestone materials, with schists and marbles. The thermotype is infratropical and the ombrotype is subhumid-humid. The presence of escarpments (cliffs) has led to the installation of edaphoxerophilous communities that must be considered as dry forest, owing to the predominance of *P. polygonus, Z. debilis, A. antillarum, Eugenia samanensis* Alain, *B. simaruba, Capparis flexuosa* L., *Ficus velutina* H. & B. ex Willd., *E. maleolens, O. dilenii, Comocladia dodonaea* (L.) Britt., *Stigmaphyllom emarginatum* (Cav.) A. L. Juss. and *C. linearis*. This area has over 60 species of flora **Figure 9**.

BA-A5. The Eastern Cordillera is the oldest range in this biogeographical territory, and has a frequent presence of limestone, karstic landscapes, tufas, alluvial deposits and foothills. It serves as a separation from the great eastern coastal plain, with sporadic intrusions of Palaeozoic slates and basalts. The thermotype ranges from infra to mesotropical; the macrobioclimate is rainy and the ombrotype is subhumid to hyperhumid. The subhumid forest with a semi-deciduous character represents the transition between the dry and ombrophilous

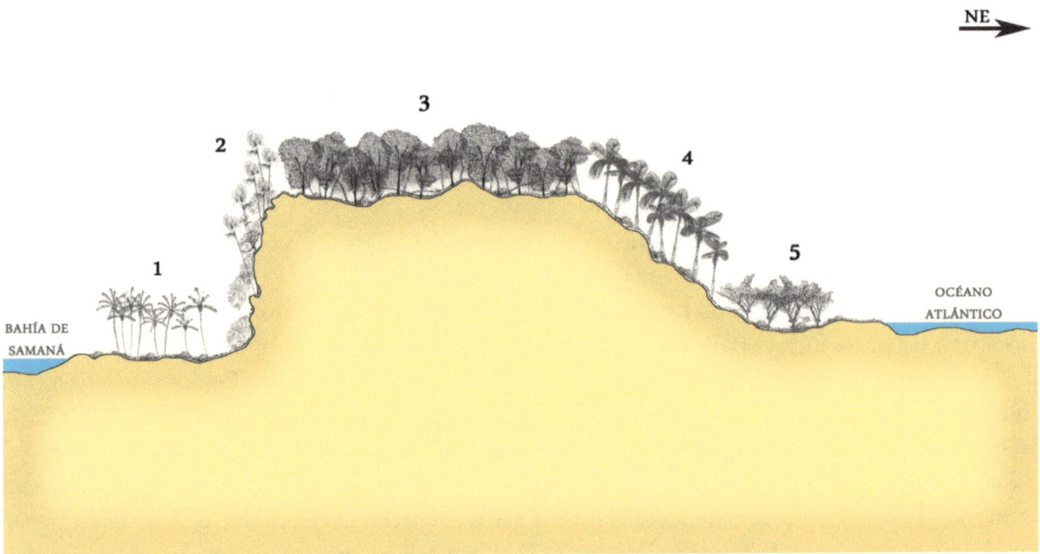

Figure 9. Vegetation catena of the Samaná Peninsula. 1. Coconut cultivation. 3. Broadleaf forest. 2. Community of *Coccothrinax gracilis* and *Bursera simaruba*. As. *Coccotrino gracili-Burseretum simarubae* [31]. 4. Cloud forest of *Prestoea montana*. 5. Forest of *Pterocarpus officinalis*. As. *Roystoneo hispaniolanae-Pterocarpetum officinalis*.

ombrotype, where there is a predominance of *S. mahagoni, C. diversifolia* and *M. toxiferum*. These formations are found primarily in the basal areas of the Eastern Cordillera, in points of contact with the Eastern Caribbean area. However, these areas are severely altered as they are used for the cultivation of cocoa, coconut and coffee, and there is a widespread presence of the cattle enclosures known as 'potreros'. This is the reason for the low rate of endemic plants, with only eight species. Above an altitude of 600 m there is a broadleaf cloud forest with a frequent presence of *D. morototoni, Inga fagifolia* (L.) Willd., *T. occidentalis, Cyathea arborea* (L.) J. E. smith, *G. guidonia, P. montana, S. virgate* and *B. plumeriana*.

BA-A6. The rainfall in Los Haitises exceeds 2000 mm, and it is home to a vegetation with a dominance of *D. arboreus, G. guidonea, S. berteriana, P. montana* and *T. occidentalis*. We propose these territories as specific biogeographical areas due to the vegetation of the 'mogotes' (steep sided residual hills) that are typical of this territory, the high rate of endemic species—with 49 endemic plants—and the resulting diversity of habitats. Its relation with A5 gives a value of $r = 0.91$ and $r = 0.71$ with A4 (Samaná) **Figure 10**.

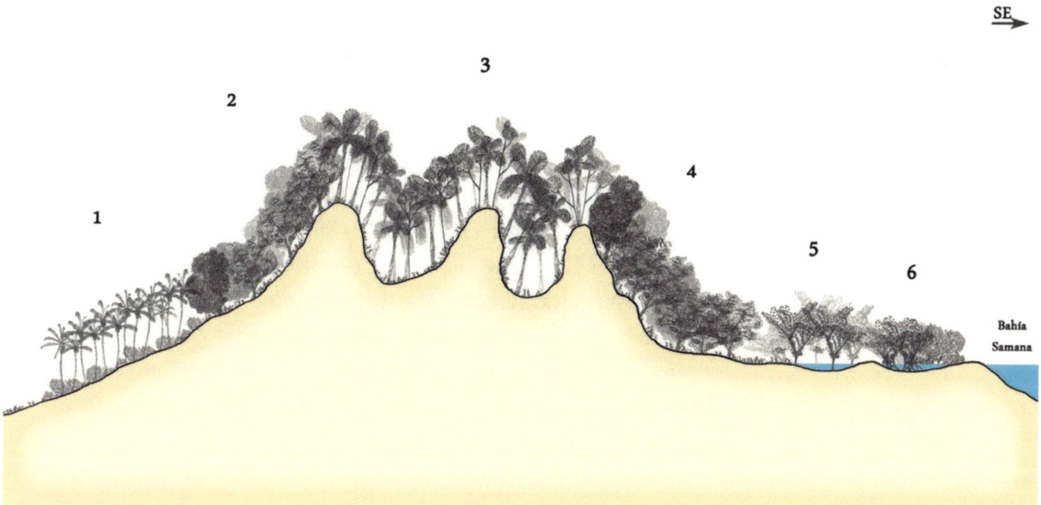

Figure 10. Vegetation catena of Los Haitises. 1. Coconut cultivation. 2. Broadleaf mahogany forest of *Swietenia mahagoni*. 3. Cloud forest of *Prestoea montana* and *Didymopanax tremulus*. 4. Broadleaf mahogany forest of *Swietenia mahagoni*. 5. Association *Roystoneo hispaniolanae-Pterocarpetum officinalis*. 6. Association *Machaerio lunati-Rhizophoretum manglis*.

5. Discussion

The island of Hispaniola is characterised by its abrupt differences in altitude—from 0 to 3175 m on Pico Duarte in the Central Cordillera—[38], the wide diversity of substrates and a pluviometric gradient that ranges from 400 to 4600 mm [35]. These three parameters, in combination with the isolation to which the various territories have been subjected, are key factors in explaining the existence of the current vegetation. For the study of this vegetation, we have established several large areas based on rainfall and temperature—dry, subhumid, humid-hyperhumid areas and high-mountain zones—as highlighted in [22, 23]. The bioclimatic analysis reveals the presence of several macrobioclimates on Hispaniola: tropical xeric, tropical pluviseasonal and tropical pluvial; all of which are reflected in different vegetation units: dry forest, subhumid broadleaf forest, cloud forest and high-mountain forest (pine forest) [35].

The dry areas have a tropical xeric macrobioclimate with a high rate of endemic species. These zones correspond closely to the study areas A3, A9 and A12 [39]. The vegetation in all the semiarid and dry areas is physiognomically very similar; it is dominated essentially by plants from the families *Agavaceae* and *Cactaceae* among others: *Lemaireocereus hystrix* (Haw.) B.&R., *Cylindropuntia caribae* (B.&R,) Kunth, *Consolea moniliformis* (L.) Haw., *Leptochloopsis virgata* (Poir.) Griseb., *Pilosocereus polygonus* (Lam.) B.& R., *Opuntia dillenii* (Fer.- Gawl) Haw., *Leptocereus weingartianus* (Hartm.) Britt. & Rose, *Acacia skleroxyla* Tuss., *Agave antillarum* Descourt. and *Pithecellobium unguis-cati* (L.) Mart. In the southwest of the island (A12), we establish two types of dry forest: first, the forest of Pedernales-Ceitillan (Procurrente de Barahona), growing on dogtooth limestone substrates. We highlight as endemic species *Melocactus pedernalensis* (Ait.) M. Mejía & R. García, *Galactia dictyophylla* Urb., *Coccoloba incrassata* Urb., *Caesalpinia domingensis* Urb. and *Guettarda stenophylla* Urb. The dry forest in area A9 with an *Io* = 2.7 has a somewhat lower rate of endemics. The most notable endemics and those which mark the difference with the dry forest of Pedernales are *Melocactus lemairei* (Monv.) Miq. *Neoabbottia paniculata* (Lam.) Britt. & Rose and *Coccotrinax spissa* Bailey. In area A3, located in the northwest of the island, there is a dry forest differentiated from the previous forests by the presence of a floristic contingent of endemic species, including *Salvia montecristina* Urb. & Ekm., *Mosiera urbaniana* Borhidi, *Croton poitaei* Urb., *Croton sidaefolius* Lam., *Guettarda tortuensis* Urb. & Ekm. and *Coccoloba buchii* Urb. The most representative plant communities in the dry areas belong to the following endemic habitats: *Lepotogono buchii-Leptochloopsietum virgatae* Cano et al. [24, 25], included in the serpentinicolous endemic alliance *Tetramicro canaliculatae-Leptochloopsion virgatae* Cano, et al. [24, 25]; *Crotono astrophori-Leptochloopsietum virgatae* Cano, et al. [24, 25], *Melocacto pedenalensi-Leptochloopsietum virgatae* Cano, et al. [24, 25], *Solano microphylli-Leptochloopsietum virgatae* Cano et al. [24, 25], included in the endemic alliance *Crotono poitaei-Leptochloopsion virgatae* Cano et al. [24, 25]; the dry forests published in Ref. [29], and the pine forests on serpentines of *Leptogono buchii-Pinetum occidentalis* Cano, Veloz & Cano-Ortiz 2011, which we include in the endemic alliance *Phyllario mummularioidi-Leptogonion buchi* Cano, Veloz, & Cano-Ortiz 2011.

Most of Hispaniola has a pluviseasonal tropical macrobioclimate and a predominantly subhumid ombrotype, with rainfall ranging from 1000 to 2000 mm and an ombrothermic index of *Io* = 3.7–4.3 (Parque Nacional del Este); *Io* = 4 (El Seibo); *Io* = 6.2 (Miches); *Io* = 5.4 (Jarabacoa) and *Io* = 5.9 (Mayaguana) (A7). The dominant vegetation in these areas is a subhumid broadleaf forest subjected to a dry season from December to April, which is why the floristic composition includes deciduous tree species due to water stress, such as *Bursera simaruba* (L.) Sarg. and *Swietenia mahagoni* (L.) Jacq., along with other species such as *Metopium toxiferum* (L.) Krug & Urb., *Krugidendron ferreum* (Vahl) Urb., *Acacia macracantha* H. & B. ex Willd., *Coccoloba diversifolia* Jacq. and *Bucida buceras* L. These formations contain important endemic elements such as the climber *Aristolochia bilobata* L. and the tree element *Melicoccus jimenezii* (Alain)

Acev. Rodr., in addition to scrubland plants such as *Lonchocarpus neurophyllus* Benth., along with the other scrubland formations that become dominant and act as dynamic substitution stages. This is the case of *Zamia debilis* L., which coexists with the endemic species *Pereskia quisqueyana* Alain and *G. ekmanii* O.E. Schulz.

When these subhumid forests are located on perforated reef limestone, the territory acts as dry owing to the intense water losses from the soil, and present the floristic elements *P. polygonus, P. unguis-cati, L. weingartianus* and *Hylocereus undatus* (Haw.) Britt. & Rose. These formations connect with the dry forest in the southwest of the island. A similar phenomenon occurs in the rocky escarpments of Samaná, where there is a widespread frequent presence of *B. simaruba, Coccothrinax gracilis* Burret, *A. antillarum, L. weingartianum* and *O. dilleni*. These habitats tend to contain deciduous species due to water stress and correspond to the associations recently proposed by us [30]: *Chrysophyllo oliviformi-Sideroxyletum salicifolii* Cano & Veloz 2012 and *Zamio debilis-Metopietum toxiferi* Cano & Veloz 2012. In dry and subhumid areas, the serpenticolous vegetation is of great interest for conservation [28].

Humid areas have a tropical pluvial macrobioclimate, and there is therefore no dry season. Rainfall exceeds 2000 mm. These humid areas tend to be located in the mountain ranges of the Northern Cordillera, Central Cordillera, Sierra de Bahoruco, Eastern Cordillera, Los Haitises and on the Samaná Peninsula, all of which concentrate the humid rainy formations, namely broadleaf ombrophilous forests whose physiognomy varies from one place to another. In the Loma La Herradura (Eastern Cordillera), the dominant plants are *Sloanea berteriana* Choisy, *Ormosia krugii* Urb., *Didymopanax morototoni* (Aubl.) Dcne. & Planch. and *Oreopanax capitatus* (Jacq.) Dcne. & Planch. Towards the stream beds, there is a presence of the sierran palm forest of *P. montana* (Grah.) Nichol, whose associated flora are *Guarea guidonia* (L.) Sleumer, *D. morototoni, Alchornea latifolia* Sw. and *Eugenia domingensis* Berg [11].

In the Central Cordillera (A16), for example, in the Ébano Verde Science Reserve, the ombrophilous forest is dominated by species from the genus *Magnolia*, which are endemic to the island: *Magnolia pallescens* Urb. & Ekm. and *Magnolia domingensis* Urb., along with the 'palo de viento' *Didymopanax tremulus* Krug & Urb., *Ocotea leucoxylon* (Sw.) Lanessan, *Persea oblongifolia* Kopp, *Cyrilla racemiflora* L., *Cecropia schreberiana* Miq. and *Dendropanax arboreus* (L.) Decne. & Planch. This forest is home to the endemic species *Myrsine nubicola* A. Liogier, *Odontadenia polyneura* (Urb.) Woods, *Marcgravia rubra* A. Liogier, *Pinguicula casabitoana* J. Jiménez and *Tabebuia vinosa* A. Gentry. As in the Loma La Herradura, the sierran palm forest of *P. montana* can be found in the most humid gorges. When these plant communities become altered and their coverage decreases, they are quickly superseded by tropical fern or herb formations of *Dicranopteris pectinata* (Willd.) Underw. and *Gleichenia bifida* (Willd.) Spreng. [9].

In the Loma Humeadora, the cloud forest of 'palo de viento' *D. tremulus* grows at an altitude of 1100–1315 m, and this species is associated with *Clusia clusioides* (Griseb.) D'Arcy, *C. racemiflora, Ocotea foeniculacea* Mez, *Lyonia alainii* W. Judd and *P. montana*. Descending to 850–1100 m on slopes with a gradient of 45–60° but with abundant litterfall that effectively retains water, and in gorges, *P. montana* becomes dominant associated with *A. latifolia, O. leucoxylon, Bombacopsis emarginata* (A. Rich.) A. Robins., *S. berteroana, Mora abbottii* Rose & Leon., *Turpinia occidentalis* (Vent.) G. Don, *Bactris plumeriana* Mart. and *Ditta maestrensis* Borhidi [8].

In the relevés taken both in the Central Cordillera and in Sierra Bahoruco, in addition to the existence of different substrates, the broadleaf forest shows clear floristic differences, with *M. pallescens* and *M. domingensis* in the Central Cordillera and *Magnolia hamorii* Howard in the Sierra de Bahoruco. The forest of *M. hamorii* and *D. tremulus* has a large number of associated endemic species such as *Lasianthus bahorucanus* Zanoni, *Psychotria guadalupensis* (DC.) Howard, *H. domingensis* Urb. *Mecranium ovatum* Cog. (local endemic), *Vriesea tuercheimii* (Mez.) L.B. Smith, *Macrocarpaea domingensis* Urb. *Cestrum daphnoides* Griseb. *Hypolepis hispaniolica* Maxon, *Columnea domingensis* (Urb.) Wiehler and *Ilex tuerckheimii* Loes. This vegetation was included in Ref. [22] in the classes *Ocoteo-Magnolietea* Borhidi & Muñiz in Borhidi, Muñiz & Del Risco 1979 and in *Weinmannio-Cyrilletea* Knapp 1964.

The study of high-mountain areas took place in the Central Cordillera (A16), crossing the mountain from Constanza to San José de Ocoa, and in the Sierra de Bahoruco (A12). From the physiognomic point of view, the plant formations sampled between 1203 m (Sierra Bahoruco) and 2383 m (Central Cordillera) are similar, corresponding to a pine forest of *Pinus occidentalis* Sw. These are territories with lower rainfall, as the sea of clouds from the trade winds originating the broadleaf forest lies beneath. The temperature may drop to 0°C in winter. The xericity and the low temperatures in the high mountains result in the presence of a pine forest of *P. occidentalis*, which in the Central Cordillera is accompanied by endemic species, with 8–10 endemic plants per sampling unit. This is also the case in the Sierra de Bahoruco, where the pine forest has an average of 20 endemic species per sampling. The endemic character of these two mountains is caused by their former isolation.

In the Central Cordillera, these forests grow on siliceous substrates and are home to a large number of endemic species such as *I. tuerckheimii, Ilex fuertesiana* (Loes.) Loes. *Garrya fadyenii* Hooker, *Mikania barahonensis* Urb., *Myrica picardae* Krug & Urb., *Rubus eggersii* Rydberb., *Tetrazygia urbaniana* (Cogn. in Urb.) Croizat ex Moscoso and *Fuchsia pringsheimii* Urb.; the endemic and specific parasitic species *P. occidentalis, Dendropemon pycnophyllus* Krug & Urb. and *Dendropemon constantiae* Krug & Urb. are of particular importance. In the understorey of this forest, there is a high frequency of the grass *Isachne rigidifolia* (Poir.) Urb., and when the pine forest is cleared, it is substituted by a formation of single-culm grasses dominated by *D. domingensis* Hack. & Pilg., which occupies large extensions above 1800 m in the Central Cordillera.

The pine forest of *P. occidentalis* growing on limestone in the Sierra de Bahoruco has a different floristic composition, in which the endemic species *Coccothrinax scoparia* Becc., *Agave intermixta* Trel., *Senecio barahonensis* Urb., *Cestrum brevifolium* Urb., *Eupatorium gabbii* Urb., *Lyonia truncatula* Urb., *Sideroxylon repens* (Urb. & Ekm.) TD. Pennington, *Cordia selleana* Urb., *Narvalina domingensis* Cass. and *Galactia rudolphiodes* (Griseb.) Benth. & Hook. *var. haitiensis* Urb. are of particular interest, along with some other endemic herbs such as *Pilea spathulifolia* Groult, *Tetramicra ekmanii* Mansf., *Artemisia domingensis* Urb., *Gnaphalium eggersii* Urban and *Polygala crucianelloides* DC. High-mountain pine forests that have been diagnosed by us as endemic habitats of Hispaniola [26] are *Dendropemom phycnophylli-Pinetum occidentalis* Cano, Veloz & Cano-Ortiz 2011 and *Cocotrino scopari-Pinetum occidentalis* Cano, Veloz & Cano-Ortiz 2011.

http://dx.doi.org/10.5772/intechopen.69081

5.1. Distribution analysis of endemic species

The study of the 19 areas in Hispaniola shows a wide distribution of endemic species, but with three nuclei of particular interest due to their high rate of endemic plants, as highlighted by the comparative treatment between the total number of endemic species present in an area and the endemic species that are exclusive to this area. There are a total of 2094 endemic species in the 19 areas, of which 1162 are exclusive. The difference between 2094 − 1162 = 932, confirming the high number of endemic species distributed all over the island. The highest concentrations are found in areas A12, A16, A13 (**Table 1**), whereas the rest of the areas have a lower number of endemic species, with a slight increase in areas A4 and A9. These areas continue to be of interest as they contain endemic species that are exclusive to the territory, and even endemic genera, as occurs in A18 and A19 (**Figure 11**).

Plots	A1	A2	A3	A4	A5	A6	A7	A8	A9	A10	A11	A12	A13	A14	A15	A16	A17	A18	A19
1	20	15	24	36	1	12	23	2	26	1	27	482	129	11	28	278	29	8	10
2	40	61	68	127	9	45	64	7	87	9	72	699	173	29	64	440	65	20	15

1. No. of exclusive endemic taxa by area. 2. Total number of endemic taxa by area

Table 1. Comparative analysis of the total endemic species in each area with the number of endemic species exclusive to that area.

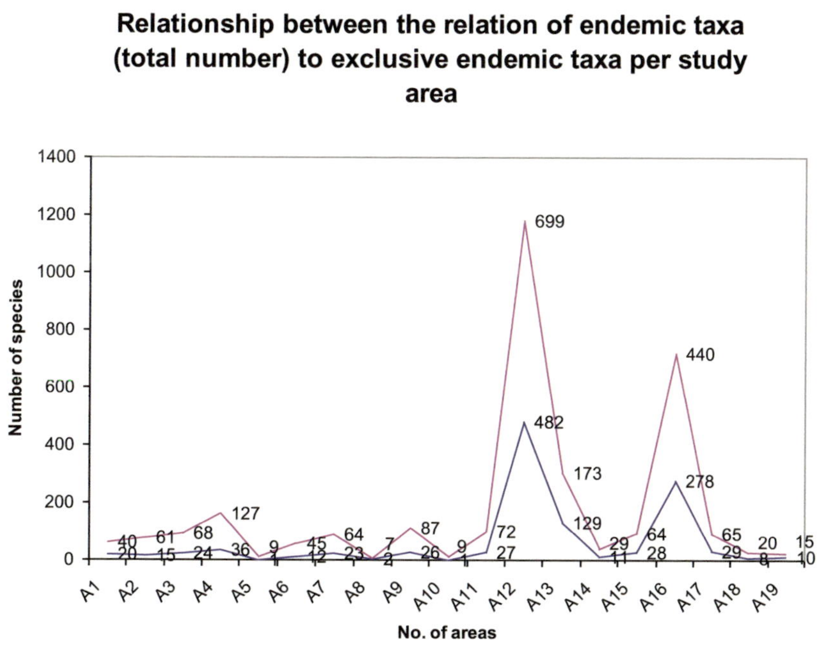

Figure 11. Ratio of total endemic species in each area to endemic species exclusive to that area.

6. Conclusions

Hispaniola has recently been elevated by us to the rank of biogeographical province [2, 22, 24], having previously been treated with the rank of biogeographical sector [4] and included in the province of the Antilles. In previous studies, we raised it to the rank of superprovince of the Central-Eastern Antilles and included Hispaniola in it, which along with a group of small neighbouring islands—Beata, Saona, Gonave and Tortuga—constitute the province of Hispaniola. In the current study, we propose a biogeographical typology with the rank of district for both countries (Dominican Republic and Republic of Haiti) in the biogeographical province of Hispaniola, in which we establish five biogeographical territories (sectors) and 19 areas (districts), in the Caribbean-Mesoamerican region. This proposal is based on geological, climatic and bioclimatic aspects and on studies of the flora and vegetation.

The high number of genera and endemic species in areas A12, A13 and A16 justifies their proposed designation as being of special interest for conservation.

Superprovince of the Central-Eastern Antilles. Province of Hispaniola. 1. **Central subprovince**. 1.1. Central BT. BA-A16. Central-Eastern. 2. **Caribbean-Atlantic subprovince**. 2.1. Bahoruco-Hottense BT. BA-A12. Bahoruco-La Selle. BA-A13. Hottense. 2.2. Neiba-Matheux-North-eastern BT. BA-A14. Neiba-Matheux. BA-A15. North-western. BA-A17. Central-Western. BA-A19. Tortuga Island. 2.3. BT Azua- San Juán-Hoya Enriquillo-Port au Prince-Artiobonite-Gonaivës. BA-A9. Azua-Sán Juán-Hoya Herniquillo. BA-A10. Central Plain. BA-A11. Port au Prince-Arbiobonite-Gonaives. BA-A18. Gonave Island. 2.4. Caribbean-Cibense BT. BA-A3. Cibao Valley. BA-A7. Eastern Caribbean. BA-A8. Yamasense. 2.5. Northern BT. BA-A1. Northern Cordillera. BA-A2. Coastal Atlantic. BA-A4. Samanense. BA-A5. Eastern. BA-A6.Haitiense **Figure 12**.

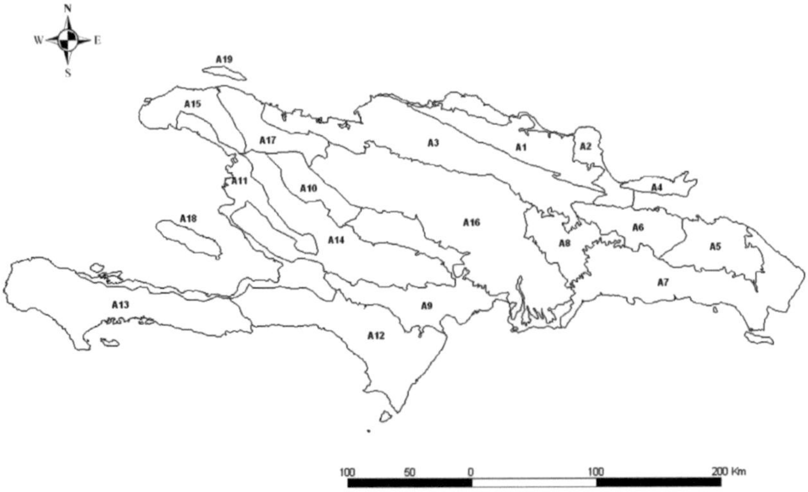

Figure 12. Map of biogeographical areas (districts) of Hispaniola. A1. Northern Cordillera. A2. Coastal-Atlantic District. A3. Cibao Valley. A4. Samanense. A5. Eastern. A6. Haitiense. A7. Eastern-Caribbean. A8. Yamasense. A9. Azua-Sán Juan-Lago Herniquillo. A10. Central Plain (Haiti). A11. Port au Prince-Ariobonite-Gonaivës. A12. Bahoruco-La Selle. A13. Hottense. A14. Neiba-Matheux. A15. Northwest Haiti. A16. Central-Eastern. A17. Central-western (Massif du Nord). A18. Gonave Island. A19 Tortuga Island.

Acknowledgements

Our most sincere thanks go to the architect Francisco Javier Quiros Higueras for designing the profiles and maps.

Author details

Ana Cano Ortiz[1], Carmelo María Musarella[2] and Eusebio Cano Carmona[1*]

*Address all correspondence to: ecano@ujaen.es

1 Department of Animal and Plant Biology and Ecology, Section of Botany, University of Jaén, Jaén, Spain

2 Departament of Agraria, University "Mediterranea" of Reggio Calabria, Reggio Calabria, Italy

References

[1] Mollat H, Wagner BM, Cepek P, Weiss W. Mapa geologic de la República Dominicana 1:250.000. Geolog. Jahrb. Hannover; 2004. p. 99

[2] Cano E, Cano-Ortiz A. Establishment of biogeographic areas by distributing endemic flora and habitats (Dominican Republic, Haiti, R.). In: Lawrence Stevens, editor. Global Advances in Biogeography. INTECH; 2012. pp. 99-118

[3] Takhtajan A. Floristic Region of the World. Transl. by Crovello TJ and ed. by Cronquist A. University of California Press; 1986. p. 554

[4] Rivas Martínez S, Sánchez Mata D, Costa M. North American boreal and western temperate forest vegetation. Syntaxonomical synopsis of the potential natural plant communities of North America, II. Itinera Geobotanica. 1999;**12**:5-326

[5] Samek V. Fitorregionalización del Caribe. Revista del Jardín Botánico Nacional (Cuba). 1988;**9**:25-38

[6] Trejo-Torres JC, Ackerman JD. Biogeography of the Antilles Based on a parsimony análisis of orchid distributions. Journal of Biogeography. 2001;**28**:775-794

[7] Borhidi A. Phytogeography and vegetation ecology of Cuba. Ed. Akadémiai Kiadó; 1991. p. 857

[8] Mejía M, Jiménez F. Flora y vegetación de Loma La Humeadora, Cordillera Central, República Dominicana. Moscosoa. 1998;**10**:10-46

[9] May TH. Respuesta de la vegetación en un calimetal de *Dicranopteris pectinata* después de un fuego, en la parte oriental de la Cordillera Central, República Dominicana. Moscosoa. 2000;**13**:113-132

[10] Zanoni TH, Mejía M, Pimentel JD, García R. La flora y vegetación de los Haitises, Republica Dominicana. Moscosoa. 1990;**6**:46-98

[11] Höner D, Jiménez F. Flora vascular y vegetación de la Loma la Herradura (Cordillera Oriental, República Dominicana). Moscosoa. 1994;**8**:65-85

[12] Guerrero A, Jiménez F, Höner D, Zanoni T. La flora y la vegetación de la Loma Barbacoa, Cordillera Central, República Dominicana. Moscosoa. 1997;**9**:84-116

[13] May TH. Fases tempranas de la sucesión en un bosque nublado de *Magnolia pallescens* después de un incendio (Loma de Casabito, Reserva Científica Ébano Verde, Cordillera Central, República Dominicana). Moscosoa. 1997;**9**:117-144

[14] May TH, Peguero B. Vegetación y flora de la Loma El Mogote, Jarabacoa, Cordillera Central, República Dominicana. Moscosoa. 2000;**11**:11-37

[15] Slocum M, Mitchell T, Zimmerman JK, Navarro L. La vegetación leñosa en helechales y bosques de ribera en la Reserva Científica de Ébano Verde, República Dominicana. Moscosoa. 2000;**11**:38-56

[16] Mejía M, García R, Jiménez F. Sub-región fitogeográfica Barbacoa-Casabito: Riqueza florística y su importancia en la conservación de la flora de la Isla Española. Moscosoa. 2000;**11**:57-106

[17] May TH. El endemismo de especies de plantas vasculares en República Dominicana, en relación con las condiciones ambientales y los factores biogeográficos. Moscosoa. 2001;**12**:60-78

[18] García R, Mejía M, Peguero B, Salazar J, Jiménez F. Flora y vegetación del Parque Natural del Este, República Dominicana. Moscosoa. 2002;**13**:22-58

[19] Veloz A, Peguero B. Flora y vegetación del Morro de Montecristi, República Dominicana. Moscosoa. 2002;**13**:81-107

[20] García R, Clase T. Flora y vegetación de la zona costera de las provincias Azua y Barahona, República Dominicana. Moscosoa. 2002;**13**:127-173

[21] Peguero B, Salazar J. Vegetación y flora de los cayos Levantado y La Farola, Bahía de Samaná, República Dominicana. Moscosoa. 2002;**13**:234-262

[22] Cano E, Veloz A, Cano-Ortiz A, Esteban FJ. Distribution of central American *Melastomataceas*: A biogeographical análisis of the Islands of the Caribbean. Acta Botanica Gallica. 2009a;**156**(4):527-558

[23] Cano E, Veloz A, Cano-Ortiz A, Esteban FJ. Analysis of *Pterocarpus officinalis* forest in the Gran Estero (Dominican Republic). Acta Botanica Gallica. 2009b;**156**(4):559-570

[24] Cano E, Veloz A, Cano-Ortiz A. Contribution to the biogeography of the Hispaniola (Dominican Republic, Haiti). Acta Botanica Gallica. 2010a;**157**(4):581-598

[25] Cano E, Veloz A, Cano-Ortiz A. The habitats of *Leptochloopsis virgata* in the Dominican Republic. Acta Botanica Gallica. 2010b;**157**(4):645-658

[26] Cano E, Veloz A, Cano-Ortiz A. Phytosociological study of the *Pinus occidentalis* forests in the Dominican Republic. Plant Biosystems. 2011;**145**(2):286-297. DOI: 10.1080/11263504.2010.547685

[27] Cano E, Cano-Ortiz A, Veloz A, Alatorre J, Otero R. Comparative analysis between the mangrove swamps of the Caribbean and those of the State of Guerrero (Mexico). Plant Biosystems. 2012. **146**(supplement):112-130. DOI: 10.1080/11263504.2012.704885

[28] Cano E, Cano-Ortiz A, Del Río S, Veloz A, Esteban FJ. A phytosociological survey of some serpentine plant communities in the Dominican Republic. Plant Biosystems. 2014;**148**(2):200-212. DOI: 10.1080/11263504.2012.760498

[29] Cano-Ortiz A, Musarella CM, Piñar JC, Spampinato G, Veloz A, Cano E. Vegetation of the dry bioclimatic areas in the Dominican Republic. Plant Biosystems. 2015;**149**(3):451-472. DOI: 10. 1080/11263504.2015.1040482

[30] Cano E, Veloz A. Contribution to the knowledge of the plant communities of the Caribbean-Cibensean in the Dominican Republic. Acta Botanica Gallica. 2012;**159**(2):201-210

[31] Cano E, Cano-Ortiz A, Veloz A. Contribution to the knowledge of the edaphoxerophilous communities of the Samana peninsula (Dominican Republic). Plant Sociology. 2015;**52**(1):3-8. DOI: 10.7338/PLS2012491/04

[32] Braun Blanquet J. Fitosociología. Bases para el studio de las comunidades vegetales. Hume H; 1979. p. 820

[33] Rivas Martínez S. Sinopsis biogeográfica, bioclimática y vegetacional de América del Norte. Fitosociología. 2004:**41**:19-52

[34] Rivas Martínez S, Loidi J. Bioclimatology of Iberian Peninsula. Itinera Geobotanica. 1999; **13**:41-47

[35] Cano E, Cano-Ortiz A, Del Río González S, Alatorre J, Veloz A. Bioclimatic map of the Dominican Republic. Plant Sociology. 2012;**49**(1):81-90. DOI: 10.7338/pls2012491/04

[36] Liogier AH. La Flora de la Española. In: Rafael Ma. editor. Jardín Botánico Nacional. Vol. I–IX. Moscoso: Santo Domingo, República Dominicana; 1996-2000

[37] Mejía M. Flora de la Española: conocimiento actual y estado de conservación. In: Actas 9° Congreso Latinoamericano de Botánica; 18-25 Junio 2006; Santo Domingo; 2006. pp. 11-12

[38] A. R. N. Atlas de los Recursos Naturales de la República Dominicana. Ed. F. Moya Pons Secretaria de Estado de Medio Ambiente y Recursos Naturales; 2004. p. 90

[39] Cano-Ortiz A, Musarella CM, Piñar JC, Pinto Gomes CJ, Cano E. Distribution patterns of endemic flora to define hotspots on Hispaniola. Systematics and Biodiversity. 2016;**14**(3):261-271. DOI: 0.1080/14772000.2015.1135195